Traffic Flow Modeling and Simulation

A Practical Approach

Traffic Flow Modeling and Simulation

A Practical Approach

Rahim (Ray) F Benekohal
University of Illinois Urbana-Champaign, USA

World Scientific

NEW JERSEY • LONDON • SINGAPORE • BEIJING • SHANGHAI • HONG KONG • TAIPEI • CHENNAI • TOKYO

Published by

World Scientific Publishing Co. Pte. Ltd.
5 Toh Tuck Link, Singapore 596224
USA office: 27 Warren Street, Suite 401-402, Hackensack, NJ 07601
UK office: 57 Shelton Street, Covent Garden, London WC2H 9HE

Library of Congress Control Number: 2026015008

British Library Cataloguing-in-Publication Data
A catalogue record for this book is available from the British Library.

TRAFFIC FLOW MODELING AND SIMULATION
A Practical Approach

ISBN 978-981-12-8677-3 (hardcover)
ISBN 978-981-12-8730-5 (paperback)
ISBN 978-981-12-8678-0 (ebook for institutions)
ISBN 978-981-12-8679-7 (ebook for individuals)

For any available supplementary material, please visit
https://www.worldscientific.com/worldscibooks/10.1142/13690#t=suppl

Desk Editors: Soundararajan Raghuraman/Steven Patt

Typeset by Stallion Press
Email: enquiries@stallionpress.com

Preface

Transportation systems are undergoing rapid transformation driven by urbanization, technological innovation, intelligent transportation systems, and the emergence of connected and automated vehicles. As these systems grow in complexity, analytical tools alone are often insufficient to capture their dynamic and stochastic nature. Traffic modeling and simulation have, therefore, become essential instruments for analysis, design, evaluation, and decision-making in modern transportation engineering.

This book on *Traffic Modeling and Simulation* (TMS) evolved from a graduate-level course that I have taught over several years. The sustained interest in the course, combined with encouragement from former students and professional colleagues, revealed a clear need for a comprehensive textbook devoted specifically to traffic flow modeling and simulation. While numerous high-quality texts address general modeling and simulation theory, there is no single volume that systematically integrates foundational modeling principles with their direct application to traffic systems. This book is intended to fill that gap.

The primary objective of this text is to provide a rigorous yet practical framework for understanding and applying traffic modeling and simulation. It emphasizes both theoretical foundations and responsible implementation. Simulation is a powerful analytical tool; when properly applied, it supports sound engineering judgment and informed policy decisions. When misapplied, it can lead to misleading conclusions and costly errors. Accordingly, this book places strong emphasis on methodological discipline, calibration, verification, validation, and statistical rigor.

The scope and organization of this book are guided by the needs of both students and practitioners. It is suitable for junior, senior, and first-year graduate courses in engineering, urban planning, computer science, operations research, management science, and related disciplines. In addition, it serves as a professional reference for transportation analysts and engineers engaged in traffic simulation studies.

This book begins with an introduction to fundamental modeling and simulation concepts, establishing the analytical framework necessary for subsequent chapters. It then presents traffic flow characteristics, including key variables such as speed, density, volume, capacity, congestion, and queue dynamics. Major traffic flow modeling approaches—macroscopic, microscopic, and mesoscopic—are examined, along with representative speed–density relationships and car following models.

Subsequent chapters address traffic simulation classifications based on flow granularity, event and process handling, simulation clock advancement, and stochastic variability. Guidance is provided on critical decisions that must be made prior to undertaking a simulation study. A comprehensive discussion of calibration, verification, and validation follows, including procedures at both micro and macro levels and statistical techniques for model assessment. The text further examines output variability analysis, batch means and replication methods, transient effects, random variate generation, comparison criteria, confidence interval estimation, sample size determination, and performance measures.

A glossary and list of abbreviations are provided at the end of this book to support clarity and consistency.

It is my hope that this book will contribute to advancing both the academic study and professional practice of traffic modeling and simulation by promoting analytical rigor, methodological transparency, and responsible application in the evolving field of transportation systems.

I would like to thank the following reviewers for their insightful comments and suggestions.

I would also like to thank my daughters, Sara and Jaclyn, for their unconditional love, support, and encouragement to finish this book.

About the Author

Benekohal is an emeritus professor of Civil and Environmental Engineering at the University of Illinois at Urbana-Champaign. He obtained his bachelor's, master's, and PhD degrees in Civil Engineering from the Ohio State University. He worked for the RKA consultant company in Tarrytown, NY, before joining the University of Illinois at Urbana-Champaign in 1987. He retired in June 2025, and currently, he is the President of Benekohal Consulting LLC. His area of research is modeling and simulation, Intelligent Transportation Systems (ITS), capacity analysis, work zone traffic management, transportation systems evaluation, railroad grade crossing operation and safety, and roadway traffic safety. He has published over 357 articles and reports in these fields. He has been the Director of the Annual Illinois Traffic Engineering and Safety Conference since 1988, was the Founder and Director of the Traffic Operations Laboratory (TOL) at UIUC, and was the former Co-Director of NEXTRANS, the Region 5 University Transportation Center. He was awarded Guest Professor of Civil Engineering, Harbin Institute of Technology, China, 2006–2009, and Honorary Professor of Traffic Engineering, Harbin Univ. of Civil Engineering and Architecture, 1996–2001. He won the Arthur M. Wellington Prize for the best paper in

Transportation, ASCE, 1993; received the Past President's Award by Illinois ITE for exceptional service to the transportation profession, 1998; received the ASCE's Forensic Engineering Outstanding Paper Award in 2016; Awarded the John LePlante Past Presidents' Award for outstanding professional service by ITE Illinois Section in 2016, and awarded the ITE Great Lakes Distinguished Member Award in 2026. He is a fellow of ASCE and ITE.

Contents

Chapter 1

Introduction to Traffic Modeling and Simulation

1.1 Transportation and Traffic Systems Simulation

The traffic–roadway environment is a complex and dynamic system. Accurately representing such a system requires tools and techniques that extend beyond simple analytical methods. Simulation provides a practical means of modeling the interactions among vehicles, drivers, roadway geometry, and control devices within a computer environment. By replicating real-world traffic flow computationally, simulation enables analysts to evaluate system behavior under a wide range of conditions.

Simulation allows the study of scenarios that may be impractical, unsafe, costly, or impossible to test in the real world. With a reliable and properly developed simulation model, alternative roadway designs, operational strategies, and traffic management policies can be evaluated efficiently and economically. However, the effective use of simulation requires a solid foundation in both traffic flow theory and the principles of modeling and simulation.

Several well-known textbooks address general modeling and simulation concepts (e.g. Banks *et al.*, 2001; Fishman, 1978; Law & Kelton, 2000; Pooch & Wall, 1992; Pritsker, 1995; Sokolowski & Banks, 2010). While these works provide valuable theoretical foundations, none is dedicated specifically to traffic flow modeling and simulation. Barceló (2010) offered detailed discussions of various traffic simulation software packages, but the material is not organized as a cohesive instructional textbook.

Over the past several decades, the use of traffic simulation models has grown substantially. Numerous commercial software packages have been developed to meet increasing demand. This trend is expected to continue with advances in Intelligent Transportation Systems (ITS) and Connected and Automated Vehicles (CAV). As transportation systems transition from exclusively conventional vehicles to mixed environments of conventional and "smart" vehicles, more sophisticated simulation tools are required to evaluate emerging operational scenarios.

When used appropriately, simulation is a powerful tool for analyzing traffic flow conditions and testing system modifications. When used improperly, however, it can lead to costly and incorrect decisions. There is, therefore, a clear need for educational resources that explain not only how to use simulation tools but also how to apply them responsibly and avoid common pitfalls.

This book presents the principles of traffic modeling and simulation and provides structured guidance for their effective application. Its objective is to equip readers with the knowledge required to model dynamic traffic systems correctly and interpret simulation results in a meaningful and statistically defensible manner.

1.2 Modeling Approaches in Transportation Analysis

When analyzing a transportation system—such as a freeway network or a weaving section—several approaches are available. One approach is experimentation with the actual system. However, real-world experimentation is often impractical, disruptive, or prohibitively expensive. An alternative is to develop a model of the system. Models may be physical or mathematical (Law & Kelton, 2000). Physical models are typically costly and difficult to construct for dynamic systems such as traffic networks. Mathematical models are more common but have limitations in representing real-world complexity. Mathematical models may be

- analytical models, which describe system behavior through equations and closed-form relationships; or
- simulation models, which replicate system behavior computationally.

Analytical models require a formal mathematical representation of system dynamics. In many real-world traffic systems, this formulation may be overly complex or impossible. Simulation provides a practical alternative for analyzing such systems.

Simulation models may be

- discrete-event models (system state changes at specific event times);
- continuous models (system variables evolve continuously over time); or
- hybrid models, combining both approaches.

Building a simulation model is only part of the overall study. The credibility of simulation results depends on careful planning, systematic validation, and rigorous statistical analysis. When properly conducted, simulation provides a powerful decision-support tool capable of evaluating complex traffic systems under a wide range of conditions. By following a structured simulation study plan, analysts can produce results that are reliable, defensible, and suitable for informed decision-making.

1.3 Simulation Study Plan

A simulation study is a structured process designed to analyze and evaluate the performance of complex traffic systems under various operating conditions. Because traffic systems involve numerous interacting components—drivers, vehicles, roadway geometry, control devices, and stochastic elements—careful planning is essential before any model is developed. A well-designed simulation study follows a logical sequence of steps to ensure that results are credible, reliable, and useful for decision-making.

The first step in any simulation study is the clear formulation of the problem. The analyst must define the study objectives, determine the scope of analysis, and identify the performance measures that will be used to evaluate system performance. These measures may include flow rate, travel time, delay, queue length, density, speed, or safety-related indicators. Clearly defining objectives at the outset prevents unnecessary modeling effort and ensures that the simulation is designed to answer specific questions rather than generate excessive or irrelevant output.

Before selecting simulation as the analysis tool, it is important to determine whether simulation is justified. Simulation should not be an automatic choice. If analytical or statistical methods can adequately evaluate the system, they should be considered first. Simulation is most appropriate when the system is too complex for closed-form analytical models, when interactions are nonlinear, when randomness plays a significant role, or when real-world experimentation is impractical, unsafe, or too costly. Additionally, simulation should only be undertaken when the analyst has sufficient understanding of the system and when reliable data are available. Without proper system knowledge and data support, a simulation model is unlikely to produce meaningful results.

Once simulation is justified, the appropriate modeling approach must be selected. Traffic simulation models vary in their level of detail:

- Microscopic models simulate individual vehicles and drivers, capturing detailed behaviors such as car-following, lane changing, acceleration, and deceleration. These models provide high realism but require substantial data and computational effort.
- Macroscopic models treat traffic as a continuous flow, focusing on aggregate variables such as average speed, flow, and density. They are computationally efficient but provide less behavioral detail.
- Mesoscopic models combine elements of both approaches, often grouping vehicles into platoons while maintaining some behavioral representation.

The choice among these approaches depends on study objectives, available data, required level of detail, and computational resources.

After selecting the modeling approach, suitable simulation software must be chosen. Software capabilities, modeling flexibility, ease of use, training requirements, cost, and known limitations must all be evaluated. In some cases, an existing program may require modification, or a new model may need to be developed to meet specific study objectives.

Reliable field data or benchmark data must then be collected. These data are essential for model calibration and validation. Data quality is critical; inaccurate or incomplete data can undermine the entire study. The analyst should identify which datasets will be used for calibration and which will be reserved for validation to avoid bias.

Establishing model credibility requires a systematic process known as Calibration, Verification, and Validation (CVV):

- Calibration involves adjusting selected model parameters so that simulation outputs closely match observed real-world conditions. Because it is not practical to adjust every parameter, attention should focus on those that strongly influence performance measures.

- Verification ensures that the model has been implemented correctly and operates according to the conceptual design. This step checks whether the coding accurately reflects the logic of the model and whether the model behaves as the experimenter assumes it would. Verification addresses the following question: *Was the model built correctly?*
- Validation determines whether the model adequately represents the real system for its intended purpose. While visual inspection can reveal obvious discrepancies, objective statistical testing provides defensible evidence regarding the agreement between simulated and observed data.

Validation typically involves comparing simulation outputs with reliable field measurements or benchmark data. Differences between simulation and field data should be evaluated to determine whether they fall within acceptable limits of normal variation. This comparison may be conducted at different levels of analysis (network, corridor, link, or intersection) and may include both microscopic measures (individual trajectories, headways, or speed profiles) and macroscopic measures (average speed, density, flow, or travel time). Validation should be performed under operating conditions similar to those for which the model will be used. A model validated only under free-flow conditions may not perform accurately under congested or stop-and-go traffic.

Once CVV is completed and confidence in the model is established, the experimental design phase begins. This phase defines the scenarios to be tested, the input variations to be introduced, and the output measures to be recorded. Because traffic simulation models often include stochastic elements, outputs may vary from one run to another even when inputs remain unchanged. This variability must be properly understood and managed. Two main sources of variability exist:

- internal variability, caused by randomness within the model (e.g. random arrival patterns or driver behavior variability);
- external variability, caused by deliberate changes in system inputs or scenarios.

To obtain reliable conclusions, multiple independent simulation runs (replications) must be performed.

Statistical methods are used to estimate the mean performance measures and construct confidence intervals. Alternatively, long simulation runs may be divided into segments using the batch means method. Transient (start-up) effects should be identified and removed to prevent biased results. Determining the appropriate number of replications is critical. Too few replications produce unreliable conclusions, while too many waste computational resources. Confidence intervals provide a quantitative measure of uncertainty and help determine when sufficient precision has been achieved.

1.4 Traffic Modeling and Simulation (TMS)

Traffic modeling and simulation represents the intersection of simulation methodology and traffic flow modeling. Effective application requires an understanding of both the theoretical and practical aspects of traffic flow and simulation techniques. In addition, users must possess knowledge of computer applications (e.g. how real-world systems are represented in code, the internal logic of simulation software, and its limitations) as well as statistics (e.g. sample size determination, data collection, experimental design, probability distributions, and data analysis).

Traffic systems are dynamic and vary over time, space, or both. They consist of multiple components whose interactions define system behavior. Because models are approximations of real-world systems (Sokolowski & Banks, 2010), they inherently have limitations. Simulation models allow repeated experimentation under varying conditions and input data, including system modifications and future "what-if" scenarios.

Simulation outputs can also be used to generate animations or visualizations of system performance. Visualization helps identify unexpected behavior and build confidence in the model, although it is not a substitute for formal validation. When used appropriately, visualization is a valuable step toward model credibility.

The intent of this book is to introduce students and practitioners to the fundamentals of traffic flow and simulation. It is designed to be practical and application-oriented and does not aim to provide an in-depth theoretical treatment of traffic flow or simulation theory.

1.5 Topics Covered

This book is divided into 10 chapters, each building upon the knowledge presented in the previous chapters.

Chapter 1—Introduction to Modeling and Simulation
Provides an overview of fundamental modeling and simulation concepts.

Chapter 2—Traffic Flow Fundamentals
Introduces basic traffic flow variables, including volume and flow rate, peak hour factor, volume vs capacity vs maximum flow rate, speed, time mean speed vs space mean speed, spacing and density, headway vs gap, fundamental traffic flow relationships, capacity and congestion, simplified queue length calculations for work zones, congestion duration and extent, and stopped vs moving queues.

Chapter 3—Traffic Flow Modeling Approaches
Covers macroscopic models (including the Greenshields model and selected speed–density relationships), microscopic models, car-following theory, car-following behavior in simulation software, relationships between macroscopic and microscopic models, and mesoscopic models.

Chapter 4—Traffic Simulation Types
Discusses traffic simulation classification based on traffic flow granularity, event and process handling, simulation clock advancement, and variability representation.

Chapter 5—What to Do before Simulation
Addresses the limitations of analytical models, justification for simulation, study goals and objectives, system definition, data collection and analysis, modeling approach and software selection, conceptual model development, calibration and validation, use of default values, selection of calibration parameters, experimental design, and output analysis.

Chapter 6—Calibration, Verification, and Validation
Presents background concepts, calibration methods for microscopic and macroscopic models, verification procedures, validation techniques, microscopic validation using speed profiles and vehicle trajectories,

macroscopic validation using traffic flow variables and fundamental relationships, comparison of simulation and field data, calibration and validation of commercial software, and examples of simulation manuals developed by state agencies.

Chapter 7—Statistical Validation Methods
Covers visual and statistical validation methods, test selection, paired and unpaired data, use of z- and t-tests, variance estimation, paired-sample t-tests, F-tests for equality of variances, chi-square tests, and Kolmogorov–Smirnov goodness-of-fit tests.

Chapter 8—Output Variability Analysis and Random Number Generation
Discusses internal and external sources of variability, batch means and replication approaches, selection of batch sizes and number of replications, initial transient effects, variance reduction techniques, correlation among batches, random number generators, linear congruential generators, and random variate generation.

Chapter 9—Comparison Criteria and Confidence Intervals
Addresses sample and population statistics, unbiased variance estimation, confidence levels and intervals, sample size determination, and performance measures for comparison.

Chapter 10—Glossary and Abbreviations
Provides the definition of the terms and abbreviations used in this book.

Exercises

1. Explain why real-world experimentation with transportation systems is often impractical. Provide three examples.
2. What are the main steps in a structured simulation study plan? Why is problem formulation critical?
3. Identify potential sources of error in a poorly planned simulation study and explain how each could affect decision-making outcomes.

References

Banks, J., Carson, J. S., Nelson, B. L., & Nicol, D. M. (2001). *Discrete-Event System Simulation* (3rd ed.). Prentice Hall, New Jersey.

Barceló, J. (Ed.) (2010). *Fundamentals of Traffic Simulation.* Springer, New York.

Fishman, G. S. (1978). *Principles of Discrete Event Simulation.* Wiley, New York.

Law, A. M. & Kelton, W. D. (2000). *Simulation Modeling and Analysis* (3rd ed.). McGraw-Hill Inc., New York.

Pooch, U. W. & Wall, J. A. (1992). *Discrete Event Simulation: A Practical Approach.* CRC Press, Boca Raton, FL.

Pritsker, A. A. B. (1995). *Introduction to Simulation and SLAM II.* Wiley, Systems Publishing Co., New York, NY.

Sokolowski, J. A. & Banks, C. M. (2010). *Modeling and Simulation Fundamentals: Theoretical Underpinnings and Practical Domains.* Wiley & Sons, New Jersey.

Chapter 2

Traffic Flow Fundamentals

Traffic flow is commonly characterized using three fundamental elements. These elements are interrelated, and knowing any two allows the third to be computed. Their mathematical relationships are discussed in the following chapter. This chapter introduces the basic concepts associated with these elements and the various ways in which they are expressed.

The three basic elements of traffic flow are as follows:

(i) flow rate or volume (denoted as q or F);
(ii) speed (denoted as u or S);
(iii) density (denoted as k or D).

Two additional elements derived from these fundamental variables are as follows:

(i) average headway (denoted as h);
(ii) average spacing (denoted as d).

2.1 Volume and Flow Rate

Traffic volume is defined as the number of units (vehicles, pedestrians, bicycles, etc.) passing a point on a roadway during a specified time interval. Although the time interval may vary, a one-hour interval is most commonly used. When vehicular traffic is counted, volume is typically expressed in vehicles per hour (vph). If the count corresponds to the busiest hour of the day, it is referred to as the peak-hour volume.

Another commonly used interval in transportation engineering is the average day. Traffic volume that represents a typical day is called average daily traffic (ADT). ADT is often estimated from short-term counts lasting from several hours to several days, but less than a full year. When traffic counts are collected continuously for more than one year and averaged, the resulting measure is annual average daily traffic (AADT). Although ADT and AADT are sometimes used interchangeably, they differ in the duration of data used to compute the average. Continuous count stations operated by state agencies are commonly used to estimate AADT.

Other volume measures include average weekday traffic (AWT), annual average weekday traffic (AAWT), and design hour volume (DHV). AWT and AAWT are based on weekday traffic counts. DHV represents the anticipated hourly volume for which a roadway is designed. For rural highways, DHV is often approximated as the 30th highest hourly volume in a year (AASHTO, 2018). For urban roadways, no single guideline exists, and DHV may correspond to a volume between the 26th and 200th highest hourly volume, depending on local conditions.

Traffic flow rate provides information similar to volume, and the two terms are sometimes used interchangeably. However, there is a technical distinction between them.

2.1.1 *Difference between Volume and Flow Rate*

The relationship between volume and flow rate is illustrated by the following example. Consider a case in which vehicle counts are recorded in 15-minute intervals over a 1.25-hour period. Hourly flow rates are computed by multiplying each 15-minute volume by four, whereas hourly volumes are obtained by summing four consecutive 15-minute counts.

Time	Volume	Flow rate (hourly)
4:45–5:00	540	540*4 = 2160
5:00–5:15	520	520*4 = 2080
5:15–5:30	600	600*4 = 2400
5:30–5:45	550	550*4 = 2200
5:45–6:00	510	510*4 = 2040

Hourly volumes are computed as the volumes in four consecutive intervals. Hourly volume between 4:45 and 5:45 is 2210 (i.e. 540 + 520 + 600 + 550 = 2210) and between 5:00 and 6:00 is 2180. It should be noted that the 15 volumes and hourly volumes come from the actual count of vehicles passing a point. However, the flow rates are computed (generated) based on the volume counts. For example, we did not count 2400 vehicles, but extrapolated it based on the 15-minute count of 600 vehicles.

2.1.2 *Peak Hour Factor*

Peak hour factor (PHF) is the ratio of the peak hour volume to max hourly flow rate within the peak hour. In the above example, the peak hour is from 4:45 to 5:45, and we counted 2210 vehicles during that hour, and max flow rate during the peak hour is 2400. Thus, the PHF is 0.92:

$$\text{Peak hour factor }(\text{PHF}) = \frac{\text{Peak hour volume}}{\text{Maximum hourly flow rate}}$$

or

$$\text{PHF} = \frac{\text{Peak hour volume}}{4\ (\text{maximum } 15-\text{minute flow rate})}$$

$$\text{PHF} = \frac{2210}{2400} = 0.92$$

PHF is usually less than 1 (typical values are between 0.85 and 0.95 for freeways). When PHF = 1, there is no volume variation within that hour. So, peak hour volume becomes equal to max hourly flow rate. When PHF is much lower than 1 (say 0.74), it implies that there is a lot of volume variation within that hour.

2.1.3 *Headway and Flow Rate*

Headway between two successive vehicles is the difference in time of passing of those vehicles over a given point on the road. Sometimes, it is referred to as time headway. Average time headway is related to the flow rate of traffic. The relationship between the average time headway (h) and flow rate (F) is

$$\text{Average headway } (h) = \frac{3600}{F}$$

where h is in seconds and F is in vph.

Example: For a flow rate of 1800 vph, average headway = $\frac{3600}{1800} = 2\,\text{sec}$.

2.1.4 *Volume vs Capacity*

Volume and capacity are two different things, even though the unit describing them could be the same. While traffic volume comes from the observed number of vehicles on the road, roadway capacity is a concept defined based on some factors. The observed volume is one of the factors considered in defining capacity. Highway Capacity Manual (HCM 7th) defines capacity as the maximum sustainable hourly flow rate under prevailing roadway, traffic, and control conditions. For highways, the capacity is expressed in terms of the number of passenger cars per hour per lane (pcphpl). When traffic is composed of cars, buses, trucks, and bikes, the non-cars are converted to equivalent passenger cars using the passenger car equivalent (PCE) factor. For example, a PCE for a large truck on a level terrain is 2, and on a rolling terrain, it is 3. It is not common, but capacity can be expressed in terms of mixed traffic in units of vph. The problem with this approach is that one needs a different capacity value for every traffic composition.

2.1.5 *Max Volume vs Capacity*

In general, the maximum observed volume is not equal to capacity, but in exceptional cases, they may be equal. If you put a sensor on a freeway and count the number of vehicles passing a point per hour, you will get the traffic volume. The maximum traffic volume your sensors measure in general is not the capacity of the highway. This is a mistake some make. Imagine the capacity of a highway is 2000 pc/hr, but max number of vehicles passing over the sensor is 1850 pc/hr. The observed 1850, even though it is the max volume at that location, is not the capacity of that location. You could process 150 additional cars per hour if the demand were there. The only time that observed volume can be equal to capacity is when enough vehicles arrive and travel at speed-at-capacity for a sustained period and flow breakdown has not occurred.

2.2 Speed

Speed tells how fast a vehicle is traveling. The common unit used in the US is feet per second (fps) or miles per hour (mph). In the metric system, it is given as meters per second (m/s) or kilometers per hour (k/h). There are several types of speeds used in the profession, such as design speed, operating speed, travel speed, and 85th percentile speed.

2.2.1 *Speed Measurement*

Speed is often measured in the field by a variety of methods, such as Doppler radar, Lidar, microwave radar, in-pavement sensors (e.g. inductive loop detectors or magnetometers), or any device that measures travel time on a given segment of the road. One can select a short distance on the roadway (say, a few hundred feet) and measure the times a vehicle enters and leaves the section. Then, find the ratio of the distance traveled over the time it took to travel that distance. This is called the "trap" method and gives speed at that spot on the road.

Two methods are often used to measure speed:

1. **Method 1 (Point speed):** Speed is measured at a point or over a short section of the road (up to a couple of hundred feet). This approach is widely used since it is an easy way to collect data. When speed is measured by Lidar, Doppler radar, or video image, it gives point speed.
2. **Method 2 (Section speed):** Speed is measured on a section of the road that may be miles long. This method is used less often because it requires more data collection efforts than Method 1. Examples of Method 2 are using aerial photos to track the vehicle over a section of road, reading license plates at two different locations, or getting an electronic signature of the vehicle at the beginning and end of the long section. This method gives the average speed of the vehicle over that relatively long section.

Depending on how the speeds are measured, average speeds are determined from a proper formula (arithmetic mean or harmonic mean formula) (Gerlough & Huber, 1975).

2.2.2 *TMS vs SMS*

Assume speeds of n vehicles are measured on a short distance of d ft (a point measurement). Let u_i be the speed of vehicle i. Let also t_i be the travel time of that vehicle on that short distance.

Two ways the speeds can be averaged are as follows:

1. Time mean speed (TMS), denoted as U_t, is the arithmetic mean of the individual speeds:

$$TMS = \frac{\sum_1^n (u_i)}{n}$$

2. Space mean speed (SMS), denoted as U_s, is the harmonic mean of the individual speeds:

$$SMS = \frac{n}{\sum_1^n (1/u_i)}$$

Another way of looking at SMS is as follows: First, find the total distance n vehicles traveled ($n*d$). Then, find the total time it took for them to travel the distance (sum of t_i). Find the ratio of $n*d$ over the sum of t_i. This ratio gives SMS:

$$SMS = \frac{nd}{\sum_1^n (t_i)}$$

Let's illustrate these averaging methods using data for four vehicles that traveled on a 300 ft section. Their travel times and speeds were measured.

Vehicle ID	Travel time (sec)	Speed (ft/sec)
1	10	30
2	7.5	40
3	6	50
4	5	60
Total	28.5	180

$$\text{TMS} = (30 + 40 + 50 + 60)/4 = 180/4 = 45 \text{ ft/sec}$$
$$\text{SMS} = 4/(1/30 + 1/40 + 1/50 + 1/60) = 42.1$$

or

$$\text{SMS} = 4*300/28.5 = 42.1 \text{ ft/sec}$$

2.2.3 *Relationships between TMS and SMS*

TMS and SMS are related to each other, and the exact relationship is (Gerlough & Huber, 1975)

$$\text{TMS} = \text{SMS} + \sigma^2_{\text{sms}}/\text{SMS}$$

where σ^2_{sms} is the variance of SMS. In general, TMS is greater than SMS, except when σ^2_{sms} is zero. This happens when the speed of all vehicles is the same. This equation is not very practical because σ^2_{sms} is not easily known.

There is an approximate relationship between SMS and TMS that is more practical to compute (Gerlough & Huber, 1975):

$$\text{SMS} \approx \text{TMS} - \sigma^2_{\text{tms}}/\text{TMS}$$

where σ^2_{tms} is the variance for TMS.

In general, TMS is about 1–2 mph higher than SMS for real-world traffic conditions. For example, from field data, the following regression equation to approximate the relationship between SMS and TMS was found (Drake *et al.*, 1967; HCM, 2000):

$$\text{SMS} = 1.026(\text{TMS}) - 1.890$$

One of the issues with this approximate equation is that when TMS is between 0 and 1.84, you get a negative value for SMS, which does not make sense. This approximate equation indicates that SMS was about 1.9 miles less than TMS in that data set.

2.3 Density

Density indicates the number of vehicles on a given length of a road. This is usually expressed in vehicles per mile (vpm) or vehicles per kilometer. Density is often computed from speed and flow rate information because measuring density directly in the field is not easy. Density can be measured in the field when you can see a long section of the road from an elevated position (e.g. airplane, helicopter, drone, or tall building).

Two density values are often used in the literature: critical density (D_c) and jam density (D_j). Critical density is also called density at capacity or optimal density. A reasonable value for critical density on freeways (assuming all passenger car traffic) is about 45 (HCM, 2000, 2022). Jam density is the density when all vehicles are stopped due to traffic congestion. It can be computed from

$$D_j = \frac{5280}{L_v + L_b}$$

where L_v is the average vehicle length and L_b is the buffer space between vehicles when they are stopped.

For example, if the average length of vehicles is 19 ft and a buffer space between vehicles in the stopped position is 7 ft, then the jam density is

$$D_j = 5280/(19 + 7) = 203 \text{ vpm}$$

A reasonable range for jam density (assuming all passenger car traffic) is around 200–250 vpm. Of course, it depends on the length of vehicles used in computing it. If traffic is composed of only tractor-trailers, jam density would be in 65–70 vpm, depending on the length assumed for a tractor-trailer.

2.3.1 *Spacing and Density*

Spacing is the distance between the front bumper of the first car and the front bumper of the second car. The relationship between average spacing (d) and density (D) is

$$\text{Average spacing } (d) = \frac{5280}{D}$$

where d is in ft and D is in vpm.

Example: For a density of 100 vpm, average spacing = $\frac{5280}{100} = 52.8\,\text{ft}$

The average spacing is occasionally called average distance headway, which can be a confusing term and is not used in this book.

2.3.2 *Relation between Average Spacing and Headway*

The average spacing and average headway are related to each other by speed:

$$d = h * u$$

where u is in ft/sec and h is in seconds.

2.3.3 *Headway vs Gap and Spacing vs Distance between Vehicles*

Looking at Figure 2.1, what is the difference between headway and gap? In computing headway, the vehicle length is included, but in computing gap, the vehicle length is not included. What is the difference between spacing and the distance between vehicles? Spacing includes the leading vehicle length, whereas the distance between vehicles (some call it buffer

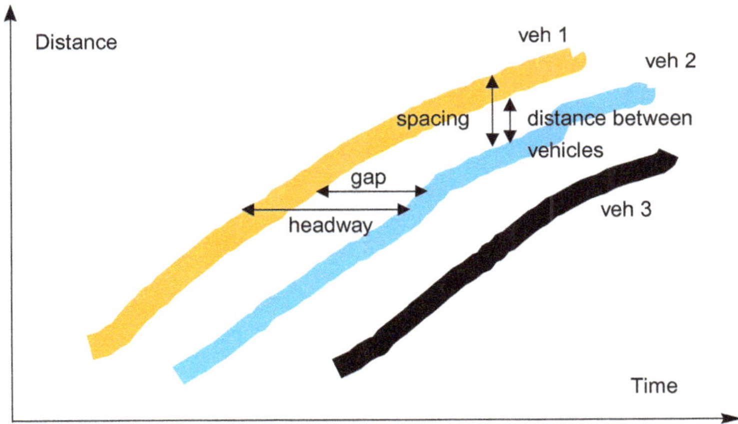

Figure 2.1. Trajectories of vehicles and illustration of headway and spacing.

or clearance) does not include vehicle length, and the distance is measured from the rear bumper of the lead vehicle to the front bumper of the following vehicle.

2.4 Fundamental Relationship

The following relationship is called the fundamental relationship of traffic flow, which will be discussed in more detail in the following chapter:

$$\text{Flow} = \text{Density} * \text{Speed}$$
$$F = D * S$$

or with the traditional notation, it is written as

$$q = k * u$$

In the fundamental relationship, which speed is used? SMS or TMS? It should be SMS, but sometimes, erroneously, TMS is used instead of SMS. This causes a small inaccuracy that is sometimes ignored.

Here is a simple example to illustrate relationships. Assume the flow rate is 1800 vph and the speed is 30 mph. Find density, average spacing, and average headway:

$$\text{Density} = \text{flow/speed} = 1800/30 = 60 \text{ vpm}$$

$$\text{Average spacing} = 5280/\text{density} = 5280/60 = 88 \text{ ft}$$

$$\text{Average headway} = 3600/\text{flow} = 3600/1800 = 2 \text{ sec}$$

Also, it should be agreed that

$$\text{Average headway} = \text{Average spacing/speed}$$

$$\text{Average headway} = (88/1.467)/(30) = 2 \text{ sec}$$

2.5 Capacity and Congestion (Queue)

What are the relations between arrival, demand, capacity, and congestion (queue)? To illustrate the relations, assume there is a construction zone on a two-lane-per-direction highway where one of the lanes is closed. Arrival volume for this work zone is the number of vehicles arriving upstream of

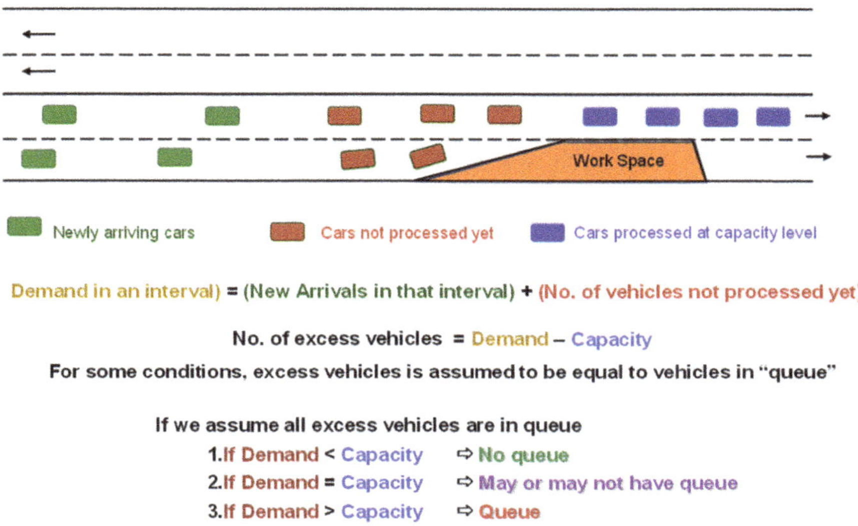

Figure 2.2. Relations between arrival, demand, capacity, and queue.

the queue and joining the back of the queue. Demand for this site is the number of arrivals added to the number of vehicles already waiting in the queue to go through the work zone. Capacity is the sustainable maximum number of vehicles that can be processed before the flow breaks down.

Congestion (queue) indicates that the number of processed vehicles is less than the demand. Obviously, when demand is less than capacity, there is no queue. If demand is greater than capacity, there will be a queue. If demand is equal to capacity, there may or may not be a queue. Figure 2.2 shows these points.

2.5.1 *Simplified Queue Length Calculation in Work Zones*

This simplified version of queue length calculation is only valid for a static queue (a stopped queue). Assume arrival volumes per hour are given, and capacity per hour (departing vehicles) is also given. Using the idea presented before, the number of excess vehicles at the end of each interval can be computed.

Number of excess vehicles = Cumulative arrival – Cumulative departure

If we assume all the excess vehicles are in queue (note that this assumption is not valid in all cases), then the number of vehicles in queue will be equal to the number of excess vehicles at the end of that time interval (Table 2.1).

It should be noted that the cumulative departure cannot be greater than the cumulative arrival. This is why the last entry in cumulative departure is set equal to the cumulative arrival (5300) for that interval. Schematically, the number of vehicles in queue is shown in Figure 2.3.

Table 2.1. Cumulative arrival and departure to find excess vehicles.

Interval	Arrival (vph)	Capacity (vph)	Cumulative Arrival	Cumulative Departure	Excess Vehicles
2–3 pm	900	900	900	900	**0**
3–4 pm	1400	900	2300	1800	**500**
4–5 pm	1400	900	3700	2700	**1000**
5–6 pm	700	900	4400	3600	**800**
6–7 pm	500	900	4900	4500	**400**
7–8 pm	400	900	5300	5300	**0**

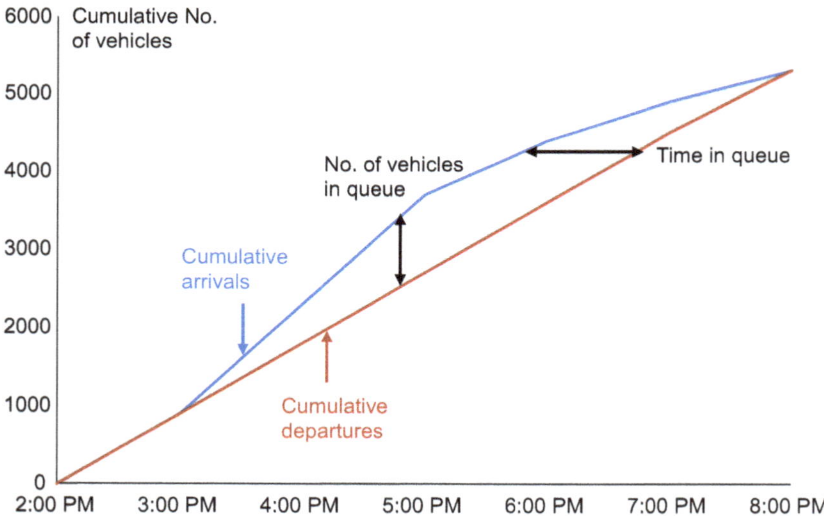

Figure 2.3. Cumulative arrival and departure plot.

2.5.2 *Congestion Duration and Extent*

Congestion (queue) can last longer than the interval it actually happened. Table 2.2 illustrates this point. Excess vehicles are obtained by subtracting cumulative departures from cumulative arrivals, and it is assumed to be equal to the number of vehicles in queue.

Assume the capacity of that road is 2240 pc/hr and the hourly arrival volume is 2210 pc/hr. Will there be any congestion (queue) even though the hourly capacity is greater than peak hour volume (2240 > 2210)? The answer is yes; there will be a queue for a period of time. Compute capacity for a 15-minute interval, 2240/4 = 560 pc per 15 minutes.

In what interval will the congestion start? It starts in the interval 5:15–5:30 because the sub-hourly flow exceeds the sub-hourly capacity. In what interval will the congestion end? It would end in the interval 5:45–6:00 toward the end of that interval. Note that there is congestion in three intervals, not just one interval.

This solution was an approximate way of finding the beginning and end of congestion. To compute the exact beginning and end of congestion, one can plot the cumulative arrival and departure curves (Figure 2.4) and find the intersection of the cumulative arrival and departure (capacity) lines. To find the exact time, the two lines intersect each other, and the horizontal axis is changed to smaller intervals (seconds). The congestion starts at 1801 sec after the beginning of the study time and ends at 4140 sec. It is assumed that arrivals and departures were uniform throughout the intervals.

Table 2.2. Cumulative arrival and departure to find the extent of queue.

Time Interval	Arriving Volume	Capacity	Cumulative Arrivals	Cumulative Departures	Excess vehicles (queue)
4:45–5:00	540	560	540	540	0
5:00–5:15	520	560	1060	1060	0
5:15–5:30	600	560	1660	1620	40
5:30–5:45	550	560	2210	2180	30
5:45–6:00	510	560	2720	2720	0

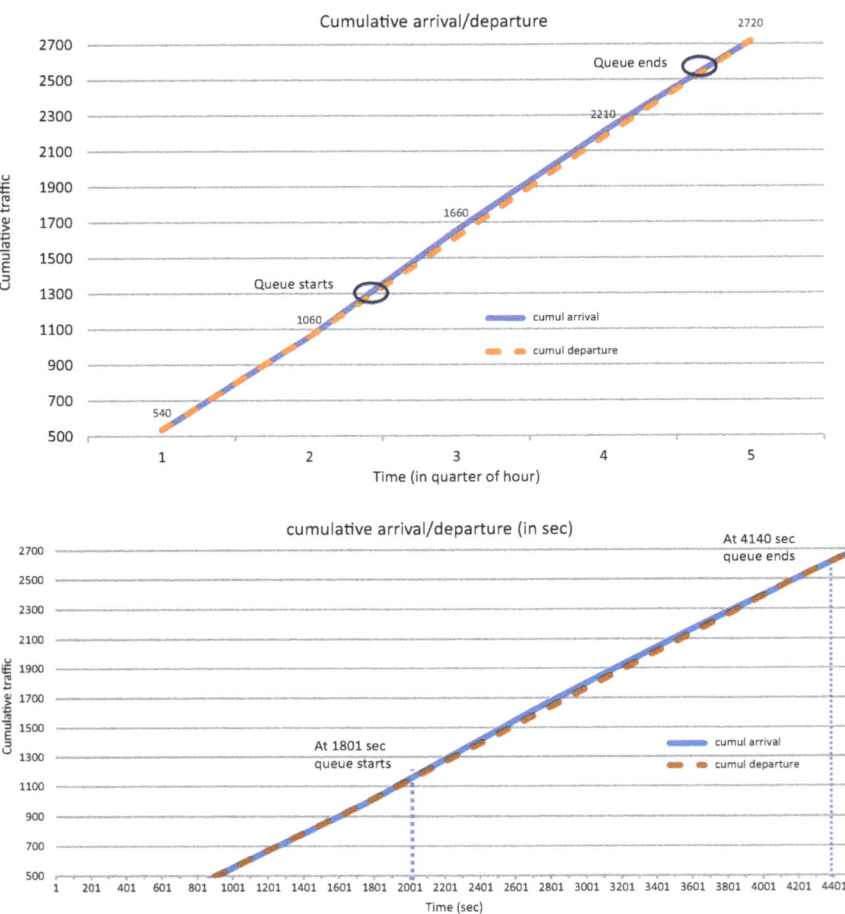

Figure 2.4. Cumulative arrival and departure to find the start and end of the queue.

2.6 Stopped Queue vs Moving Queue

There are two types of queues you may encounter in a construction zone:

(1) stopped queue (static queue);
(2) moving queue (dynamic queue).

In a stopped queue (also called a static queue), virtually all vehicles are stopped. In a moving queue, vehicles are not stopped but moving forward, and the queue is growing and/or shrinking. Queue length

Storage Space for Stopped Queue versus Moving Queue

- What would be required storage space for 1000 stopped vehicles?

1000 999 • • • • • • 4 3 2 1

Storage space = No. of vehicles∗spacing in stopped queue
= 1000∗25 = 25000 ft ≈ 4.73 miles

- What would be required storage space for 1000 moving vehicles?

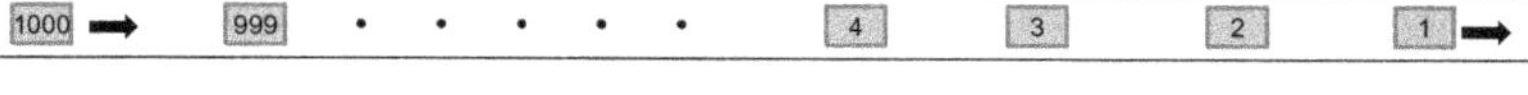

Storage space = No. of vehicles∗spacing in moving queue
For flow rate = 1800 & speed = 30 mph, Average spacing is 88 ft
Storage space = 1000∗88 = 88000 ft ≈ 16.67 miles
What if we assume spacing is 66 ft
Storage space = 1000∗66 = 66000 ft ≈ 12.5 miles

Figure 2.5. Length estimations for stopped and moving queues.

calculation for a static queue is relatively simple, but it gets complicated for a moving queue. Often, the queue we encounter on a highway bottleneck (e.g. work zone and accident site) is a moving queue, and the queue at signalized intersections is a static queue.

Figure 2.5 shows how to compute the queue length for stopped and moving queues if we assume the number of vehicles in the queue is known.

For a stopped queue, it is relatively easy to determine/assume a reasonable spacing between vehicles. The spacing used will depend on the traffic composition, and the average vehicle length should be the weighted average of the length of all vehicles. In the above example, an average vehicle length of 19 ft and a buffer space of 6 was used. So, the average spacing is 19 + 6 = 25 ft for the stopped queue. Thus, the stopped queue length was 1000∗25 = 25,000 ft (4.73 miles).

However, for the moving queue, the average spacing will be a function of the speed of vehicles, and this is not easy to determine. If we assume 88 ft spacing vs 66 ft spacing, the queue length estimated would be substantially different. The queue length would be 88,000 ft if 88 ft spacing is used and 66000 ft if 66 ft spacing is used. Determining spacing for a moving queue is complicated because a moving queue is a dynamic queue, and it may shrink (expand) from front, back, or both ends. For a moving queue, we cannot give a "back of the envelope" type simple calculation similar to the one we discussed for a stopped queue. It would require more calculation and often use software/spreadsheet. One of the software programs to calculate the moving queue length in a work zone is WorkZoneQ (Benekohal *et al.*, 2013).

Exercises

Part 1. Questions

1. Is AADT a more stable volume than ADT? Why?
2. Define the three fundamental traffic flow variables. Explain how any two can be used to compute the third. Which speed (TMS or SMS) should be used in the fundamental relationship? Why?
3. Explain the difference between traffic volume and flow rate. Under what condition would they be equal?
4. Define Peak Hour Factor (PHF). What does PHF = 1 imply? What does PHF = 0.75 indicate?
5. Define Time Mean Speed (TMS) and Space Mean Speed (SMS). Why is SMS usually less than TMS? When would they be equal?
6. Explain why the maximum observed volume is generally not equal to capacity.
7. Explain the difference between a stopped queue and a moving queue. Why is the moving queue length harder to estimate?

Part 2. Problems

Problem 1. A freeway lane carries 2100 vehicles per hour at an average speed of 42 mph. Compute density (vpm), average spacing (ft), and average headway (sec). Verify that $d = u \times h$.

Problem 2 (Peak Hour Factor). 15-minute volumes over a 1.5-hour period are as follows: 480, 600, 620, 500, 560, and 510 vehicles. (a) Compute the peak hour volume (maximum total over any four consecutive 15-minute intervals). (b) Compute the maximum hourly flow rate within the peak hour (4 × maximum 15-minute volume). (c) Compute PHF.

Problem 3. Five vehicles travel a 500-ft section at speeds 25, 35, 40, 50, and 60 ft/sec. Compute TMS and SMS.

Problem 4. Average vehicle length is 20 ft, and stopped buffer spacing is 8 ft. Compute jam density. Then also compute jam density if trucks average 65 ft in length and 10 ft in spacing.

Problem 5. A work zone capacity is 1200 vph. Arrival volumes by hour are 1000, 1600, 1800, 900, and 700 vph. Compute cumulative arrivals, cumulative departures, maximum queue, and when congestion starts and ends.

Problem 6. If 850 vehicles are in a stopped queue with an average spacing of 25 ft, compute the queue length in miles. If the moving spacing is 95 ft, compute the moving queue length.

Problem 7. The following are 15-minute traffic volume data collected in the field. Assume arrival and departure rates are constants, and any excess vehicle is in queue.

Time	**Volume (vehicles per 15 min)**
7:30–7:45	500
7:45–8:00	530
8:00–8:15	540
8:15–8:30	505
8:30–8:45	480
8:45–9:00	475

(a) Find hourly volumes and flow rates.
(b) Find peak hour and PHF.
(c) Plot the cumulative number of arrivals and departures assuming the departure rate is 2040 vehicles per hour per lane (vphpl).
(d) If the capacity of the lane is 2040 vphpl, what is the maximum number of vehicles in queue?
(e) What interval would the queue start and end?
(f) When (accurate to seconds) is the earliest time there will be at least one vehicle in queue (assuming arrival and departure are uniform)?
(g) When (accurate to seconds) would the congestion end (no vehicle in queue), assuming arrival and departure are uniform?

Problem 8. The average space mean speed of a traffic stream is 52 mph, and the average headway is 1.92 sec. Find flow rate, density, and spacing for the traffic. If you assume that density at capacity is 45 vpm, is the traffic in congested or uncongested state?

Problem 9. The time vehicles took to travel a 100 ft section of a road was measured. The measurements lasted for 15 minutes, and the travel times were grouped as follows:

Travel Time (sec)	Frequency of observing this time
2.2	37
2.3	35
2.4	44
2.7	32
2.8	25
3.1	17

Find time mean speed, space mean speed, flow, and density.

References

American Association of State Highway and Transportation Officials. (2018). *A Policy on Geometric Design of Highways and Streets* (7th ed.).

Benekohal, R. F., Ramezani, H., & Avrenli, K. A. (2013). *WorkZoneQ User Guide for Two-Lane Freeway Work Zones* (Report No. FHWA-ICT-13-019). University of Illinois at Urbana–Champaign. https://ideals.illinois.edu.

Drake, J. S., Schofer, J. L., & May, A. D. (1967). A statistical analysis of speed-density hypotheses. *Highway Research Record, 154*, 53–87. Transportation Research Board.

Gerlough, D. L. & Huber, M. J. (1975). *Traffic Flow Theory: A Monograph* (Special Report 165). Transportation Research Board, National Academies.

HCM. (2000). *Highway Capacity Manual 2000*. National Academies Press.

HCM. (2022). *Highway Capacity Manual* (7th ed.). National Academies Press.

Chapter 3

Traffic Flow Modeling Approaches

Three approaches, broadly speaking, may be taken to develop mathematical models of traffic flow:

1. macroscopic approach, resulting in macroscopic models;
2. microscopic approach, resulting in microscopic models;
3. mesoscopic approach, resulting in mesoscopic models.

Mathematical models of traffic flow are expressions used to represent real-world traffic flow in the form of equations and/or relationships. The models are used to represent the traffic flow in computer environments for analytical or simulation studies. Each approach is presented, and the main advantages and disadvantages of them are discussed.

3.1 Macroscopic Approach

In the macroscopic approach (that yields the macroscopic models), the traffic flow is represented in terms of aggregate characteristics of the traffic stream (macroscopic variables). Traffic flow is described in terms of the average speed of the traffic stream, traffic volume, and traffic density. Unlike the microscopic approach, the macroscopic approach does not consider the individual characteristics of the vehicle-driver units. In the macroscopic approach, traffic flow is assumed to resemble the flow of a continuous fluid (like the flow of water in a channel). Hydrodynamic analogy was originally suggested by Lighthill & Whitham (1955), as well as by Richards (1956) independently. This approach is sometimes referred to as the LWR approach, even though Richards did not work with the other two authors. These models use partial differential equations (PDEs) to simulate traffic dynamics.

In general, the macroscopic models assume that the flow of traffic is analogous to the flow of compressible continuum media. The continuum flow models can be grouped into two categories based on the speed adaptation assumption: (1) First-order models (simple order model), which include LWR, and cell transmission models (CMTs) introduced by Daganzo (1994); CMT is considered as a discretized version of the LWR approach. This category assumes that speed is an explicit function of density, and drivers adjust their speed instantaneously to the density around them. In other words, the traffic state (defined by speed and density) jumps from one state to another state without transitioning. (2) Higher-order models, which attempted to address the issue of "jump" in states of traffic in the first-order models. A high-order model was introduced by Payne (1971) that added a momentum (velocity) equation to incorporate two human factor elements of relaxation and anticipation. The relaxation term considers that drivers take time to adjust to their desired speed. The anticipation term considers that drivers look ahead and reduce their speed if the density increases. Whitham independently proposed a higher-order model that incorporated the dynamics of driver behavior (Whitham,1974). Some people refer to these models as the P-W model. Adding the "momentum" term made the high-order models more realistic, but it brought unreasonable wave propagation issues as discussed by Daganzo (1995). Then, Aw & Rascle (2000) and independently Zhang (2002) modified the anticipation term to address the wave propagation issues. Some people refer to these works as ARZ models.

The three variables to describe traffic flow are speed (u), density (k), and volume (q). These variables change over time (t) and space (x). They are related to each other with the following relationship, which is called the fundamental relationship of traffic flow:

$$q = k * u$$

This equation says flow (or volume) is equal to speed times density. Considering the changes over space and time, it may be written as

$$q(x, t) = k(x, t) * u(x, t)$$

Another equation that relates the three variables is the continuity equation that is expressed as

$$\frac{\partial q}{\partial x} + \frac{\partial k}{\partial t} = 0$$

To solve this equation, we hypothesize that in steady state traffic, flow is a function of density $q = q(k)$ and speed is a function of density $u = u(k)$. The continuity equation implies that when traffic is in a steady state condition (equilibrium condition), the number of vehicles is conserved between two points on the road where there are no entrances or exits in between. For example, if there were 9 vehicles already on a section of a road and 48 new vehicles entered the section and 46 vehicles exited from the section, the number of vehicles remaining on the section must be 11. For the situations where there are entrances and/or exits on the section and vehicles are entering and/or leaving the section, a term can be added to the right-hand side of the continuity equation.

3.1.1 *Greenshields Model*

A historic and popular macroscopic model is the Greenshields model. In the Greenshields model, it is assumed that speed is a linear function of density. Greenshields proposed this model in 1937 based on data collected on rural highways (not freeways) in Ohio.

The general form of the Greenshields model is

$$u = a - bk$$

where u is the speed and k is the density of the traffic stream. Applying boundary conditions, one can derive the Greenshields model. First, boundary condition is when the density is near zero ($k = 0$), the speed is called the free flow speed (u_f). Thus, $u_f = a$. The second boundary condition is when speed is zero due to traffic congestion ($u = 0$) and the density represents jam density k_j. Thus, $k_j = a/b = u_f/b$. So, $b = u_f/k_j$.

Hence, the Greenshields model is

$$u = u_f - (u_f/k_j) * k = u_f(1 - k/k_j)$$

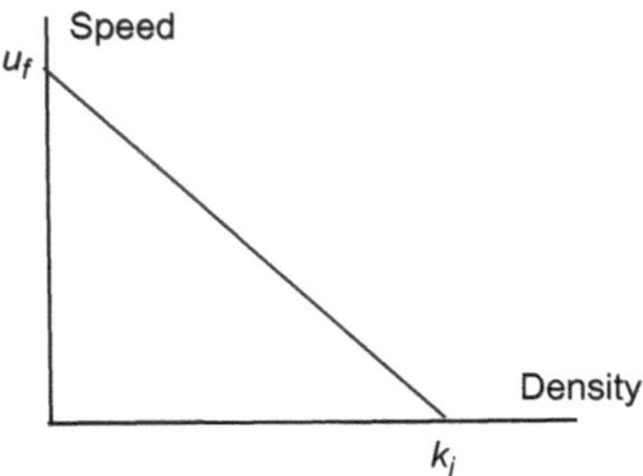

Derivation of the Greenshields models is presented in the following. Relationship between speed and density is

$$u = u_f\left(1 - \left(\frac{k}{k_j}\right)\right)$$

Substituting the above value for u in $q = u * k$, one gets a relation between flow and density:

$$q = u_f * k - k^2\left(\frac{u_f}{k_j}\right)$$

This is the equation of an upside-down parabola.

Capacity, q_c, and density at capacity, k_c, are shown.

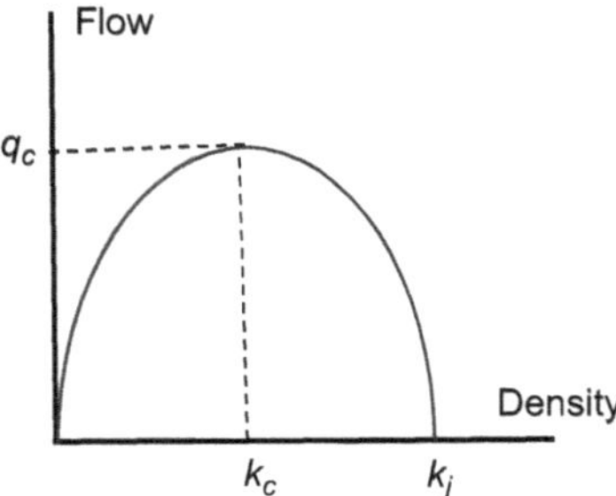

Find the derivative of q with respect to k and set it equal to zero to find the density at maximum flow, k_c:

$$\frac{dq}{dk} = u_f - 2k\left(\frac{u_f}{k_j}\right)$$

$$u_f - 2k\left(\frac{u_f}{k_j}\right) = 0$$

So, density at max flow (k_c) is

$$k_c = \frac{k_j}{2}$$

k_c is also called critical density or density at capacity. So, in the Greenshields model, the critical density is one-half of the jam density. Thus, the parabola is symmetric around *y*-axis.

Since $k = q/u$, substitution of q/u for the k in this equation $u = u_f \left(1 - \left(\frac{k}{k_j}\right)\right)$ would yield

$$u = u_f\left(1 - \left(\frac{q}{u * k_j}\right)\right)$$

Simplify it,

$$u^2 = u_f * u - \left(\frac{u_f}{k_j}\right) * q$$

This is the equation of a sideways parabola.

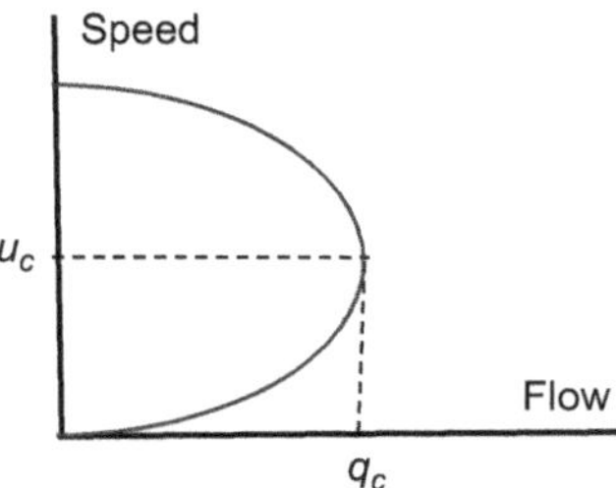

Taking the derivative of q with respect to u

$$2u = u_f - \left(\frac{u_f}{k_j}\right)\frac{dq}{du}$$

Set $\frac{dq}{du}$ equal to zero to find the speed at max flow u_c:

$$2u = u_f - \left(\frac{u_f}{k_j}\right) * 0$$

So,

$$u_c = \frac{u_f}{2}$$

This means the speed at max flow is one-half of the free flow speed. Speed at max flow, u_c, is also called speed at capacity or critical speed.

Thus, the max flow in the Greenshields model when speed is $\frac{u_f}{2}$ and density is $\frac{k_j}{2}$ is

$$q_c = u_c * k_c = \frac{u_f}{2} * \frac{k_j}{2} = \frac{u_f * k_j}{4}$$

3.1.2 *Issues with Greenshields' Model*

The previous equation showed that max flow is when traffic is at the critical speed and critical density. Let's use an example to see if the Greenshields model yields reasonable values.

Let's assume reasonable values for speed and density for a traffic condition on a rural highway similar to the one in Ohio in the mid-1930s. Reasonable values for u_f and k_j would be $u_f = 36$ mph and $k_j = 180$ vpm. Thus, the max flow would be

$$q_c = \frac{u_f * k_j}{4} = \frac{36 * 180}{4} = 1620 \text{ vph}$$

Does this seem reasonable value for capacity? Yes, it does seem reasonable for the assumed traffic conditions.

However, there is a problem with the Greenshields model when it is applied to today's freeways. For today's freeway traffic conditions, what are reasonable values for u_f and k_j? If we assume $u_f = 76$ mph and $k_j =$ 200 vpm, then

$$q_c = \frac{76 * 200}{4} = 3800 \text{ vph}$$

Does this seem a reasonable value for the capacity of a single lane of an interstate freeway? No, it is too high because it corresponds to a headway of less than 0.95 sec (3600/3800), and that is not a reasonable average headway. A reasonable headway for freeway traffic would be 1.5–2 sec.

Did Greenshields collect data on freeways? No, it was not on a freeway. Can a linear relation be adequate to represent freeway traffic? It does not seem that it can because recent data from freeways show that the speed–density relationship is not linear. So, researchers have looked for other models that would represent freeway traffic better than the Greenshields model (Drake *et al.*, 1967). Several popular speed–density models are presented here.

3.1.2.1 *Example for Greenshields Model*

Assume the following relationship between speed (u in mph) and density (k in vpm) was obtained for freeway traffic, $u = 75.4 - 0.344k$.

a. Plot speed–density graph.

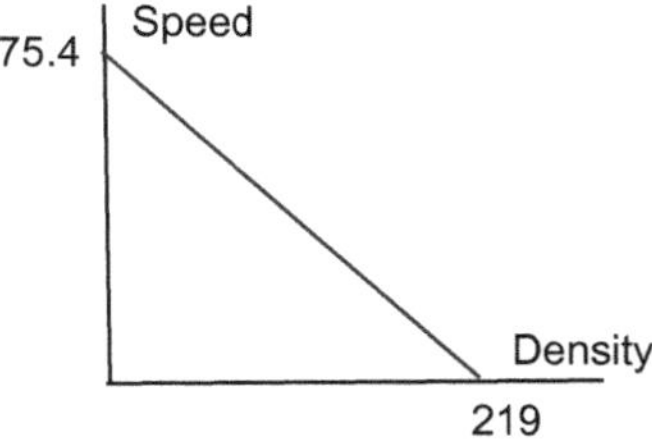

b. Find speed–flow relationship and plot it:

$$u = 75.4 - 0.344k$$
$$k = (u - 75.4)/0.344$$
$$q = k * u = u(u - 75.4)/0.344$$
$$q = (u^2 - 75.4u)/0.344$$

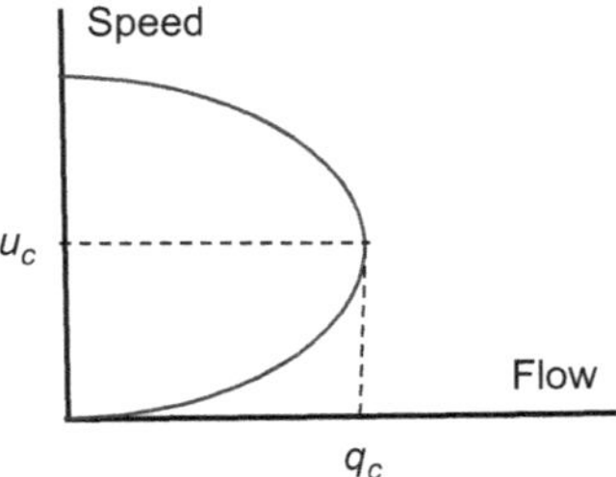

c. Find density–flow relationship and plot it:

$$q = k * u = k(75.4 - 0.344k) = 75.4k - 0.344k^2$$

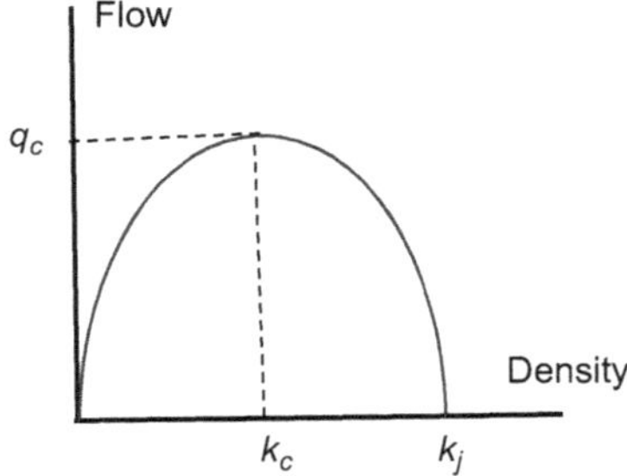

d. What is free flow speed?

Free flow speed is 75.4 mph

e. What is speed at capacity?

Speed at capacity is $u_c = u_f/2 = 75.4/2 = 37.7$ mph

f. What is jam density?

To find jam density, set speed equal to zero:

$$0 = 75.4 - 0.344k_j$$
$$k_j = 75.4/0.344 = 219 \text{ vpm}$$

g. What is critical density (density at capacity)?

$$k_c = k_j / 2 = 219/2 = 109.5 \text{ vpm}$$

h. What is max flow (capacity)?

$$q_c = u_c * k_c = (109.5)(37.7) = 4128 \text{ vph}$$

3.1.3 *Selected Speed Density Models*

There are numerous speed–density models proposed by various researchers. Publications such as Drake *et al.* (1967), May (1990), and Gerlough *et al.* (1975) discuss these models in more detail. Here, several selected ones are presented (see Figure 3.1) to introduce the readers to a wide field of traffic flow modeling.

3.1.3.1 *Single-Regime Linear Model*

Linear single regime models, often called the Greenshields model, fit a single line through the entire range of density data. The advantage of Greenshields models is that they are a simple model that yields free flow and jam density. The disadvantage of it is that a single linear equation may not fit congested and uncongested traffic conditions. To overcome the limitation of a single line for the entire data set, researchers have tried models such as two or more lines to cover different congestion levels more accurately.

3.1.3.2 *Two-Regime Linear Model*

A two-regime linear model has one equation for uncongested conditions and another equation for congested conditions. The advantage of this model is that the equations fit the congested and uncongested data closer than the single regime model. The main disadvantage of this model is the discontinuity of it that requires finding a break point that has a meaning in terms of traffic congestion, and at the same time, the lines fit the data well, and it does not create a vertical gap between the two models.

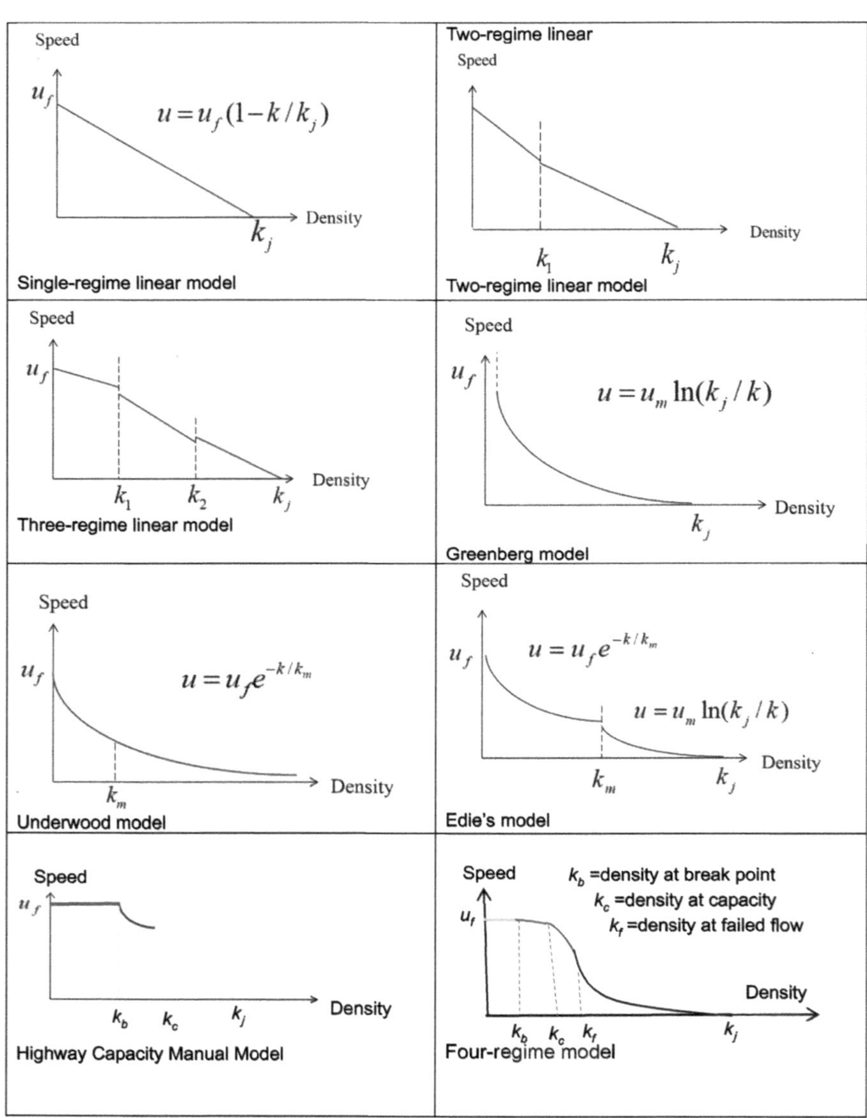

Figure 3.1. Speed–density relationships for selected models.

3.1.3.3 *Three-Regime Linear Model*

The three-regime linear model has one equation for uncongested, one equation for transition, and one for congested conditions. The advantage of this model is that the equations fit the congested, transition, and uncongested data closer than the single regime and two-regime models. The main disadvantage of this model is the two discontinuities it has that require not only finding two break points that have meanings in terms of traffic congestion but also finding equations that provide consistent slope and do not create vertical gaps between adjacent models.

3.1.3.4 *Greenburg Model*

Greenberg proposed a logarithmic model and found a good agreement between his model and field data from congested tunnels in the New York area. In the Greenberg model, U_m is a model parameter, and it has to be specified; thus, it determines the other characteristics. The advantage of the Greenberg model is that it fits the congested data well and shows a long tail for the curve toward the jam density. However, the disadvantage of this model is that it does not give a reasonable free flow speed, unless the model is truncated at a certain speed level.

3.1.3.5 *Underwood Model*

Underwood (1961) proposed an exponential form model. The advantage of this model is that it represents the uncongested part well and gives a reasonable free flow speed. However, the disadvantage of it is that it does not give a reasonable jam density value, unless it is truncated at a certain density level.

3.1.3.6 *Edie Model*

A two-regime model with one nonlinear equation for uncongested conditions and another nonlinear equation for congested conditions was proposed by Edie (1961). The advantage of this model is that the equations fit the congested and uncongested data closer than the linear regime model. The main disadvantage of this model is the discontinuity between the two models that may create a gap between the two models.

3.1.3.7 *Highway Capacity Manual Model*

Highway Capacity Manual is based on recent freeway data. Speed remains constant until traffic density reaches a value of k_b, and after that, it has a curvilinear relationship with density. The HCM model does not cover the entire density range. It is limited to traffic volumes that are under capacity, and thus, it does not offer a model for congested conditions. k_b represents the density at the break point where the horizontal part of speed–density model ends. Beyond the break point, the relationship becomes curvilinear and ends at critical density (k_c). HCM does not have a model for congested traffic conditions.

3.1.3.8 *Four-Regime Model*

A four-regime model for freeway construction zone traffic was developed by Avrenli *et al.* (2012). The four regime are as follows: (a) free flow regime (below 900 pc/hr of flow, below 15 pc/mile of density; speed is almost constant), (b) uncongested transition regime (900–1850 pc/hr of flow, 15–35 pc/mile of density), (c) congested transition regime (from 1850 to 1300 pc/hr of flow, 35–52 pc/mile of density), and (d) congested regime (below 1300 pc/hr of flow, 52 to the jam density of 271 pc/mile). To make the speed–flow curve smooth, they fit 4th degree splines to the transition regime curves. This provided a smooth change between the regimes. Their model covers the entire range of density from free flow, uncongested near capacity, congested near capacity, and flow conditions.

3.2 Microscopic Approach

In a microscopic approach, traffic flow is composed of individual entities (e.g. pedestrians, bikes, cars, and trucks) and each entity has its own characteristics. These entities are linked to each other by logical relationships, and they are moved in a simulated environment based on some rules. The behavior of the one entity in the front (leader) will influence the behavior of the entity behind it (follower) if they are close to each other. In the microscopic approach (models), traffic flow is expressed in terms of characteristics that describe individual entity such as the reaction time of a driver, the desired speed of a driver, the walking speed of a passenger, and the length of a vehicle. These unique characteristics define those entities. An entity can be a driver or a vehicle or a combination of the two. Often, a driver is assigned to a vehicle, and that combination becomes one driver–vehicle unit. Thus, traffic flow is composed of these individual entities (units) that move along a roadway environment that has many segments that have unique characteristics.

The entities are moved longitudinally and laterally through the simulated roadway system. For each entity (unit), the unique characteristics are assigned when the entities are generated in the simulation environment. For example, when a vehicle enters a roadway system, it is given certain characteristics such as length, width, acceleration/deceleration, desired speed, headway, aggressiveness, reaction time, and gap acceptance. These units are moved longitudinally by a logic called car following (CF) logic or CF model. The units are moved laterally by another logic called lane changing (LC) model. The LC models work in harmony with CF to make lateral and longitudinal movement as smooth and reasonable as possible.

3.2.1 *Car Following Models*

A car following model or a car following logic is a mathematical expression that describes how a vehicle is longitudinally moved in a virtual environment. This expression tries to mimic the way vehicles move in real world. A car following model describes how a follower reacts to its leader's maneuvers. A pair of leader–follower is used to describe the relations.

CF models are expressed in the form of a stimulus–response equation (Gerlough *et al.*, 1975):

$$\text{Response} = \text{Sensitivity} * \text{Stimulus}$$

The response term represents the acceleration/deceleration that the follower utilizes. The sensitivity term represents how sensitive the follower is to the changes in the movements of the leader. The stimulus is often the speed difference between leader–follower pairs.

Car following models are grouped into two categories: linear and nonlinear (Gerlough *et al.*, 1975). In a linear car following model, the sensitivity of the follower is assumed to be constant. In a nonlinear CF, the sensitivity of the follower is not constant but depends on factors such as the speed of the follower or the spacing between the lead–follower, or a combination of these factors.

3.2.1.1 *Example of Car Following Models*

Typically, the notations used in CF models are $x, \dot{x}, \ddot{x}$. For example, $x_n(t)$ shows the position (or location) of vehicle n at time t where x is the position (location), $\dot{x}$ is the speed, and $\ddot{x}$ is the acceleration/deceleration rate.

Variables At time *t*	**(*n* + 1)th car (follower)**	***n*th car (leader)**
Position	$x_{n+1}(t)$	$x_n(t)$
Speed	$\dot{x}_{n+1}(t)$	$\dot{x}_n(t)$
Acceleration	$\ddot{x}_{n+1}(t)$	$\ddot{x}_n(t)$

Car following models vary from simple to more complex logic. Car following studies were pioneered by Pipes (1953), Herman *et al.* (1959), Gazis *et al.* (1959, 1961), and Herman and Potts (1961). Kometani and Sasaki (1961), Herman and Rothery (1965). Other researchers such as Drake *et al.* (1967), May *et al.* (1967), Treiterer (1967), and Treiterer and Myers (1974) examined how well the car following models fit the field data from freeways. The work by Gazis, Heman, and Rothery (GHR) was conducted in General Motors' facilities near Detroit, MI, and some authors (e.g. May, 1990) refer to these models as GM1 through GM5 models. For more details on the development of car following models, consult the following references (May, 1990; Gerlough *et al.*, 1975; Benekohal, 1989; Aycin *et al.*, 1998).

3.2.1.2 *Simplest CF Model*

The follower moves in a way that the spacing between the pair of vehicles is a constant number. It may be expressed as

$$x_n(t) - x_{n+1}(t) = \text{constant}$$

Based on this model, the following vehicle keeps a constant distance from the lead vehicle. This is a very simple model but is not a good one because it is not close to what goes on in real traffic conditions. Why is it not a good model? It is not a good model because the spacing between the pair of vehicles remains constant regardless of how far they are from each other, and at what speeds they are traveling. If the lead car is 150 ft or 2000 ft ahead of the follower, the following car keeps the same spacing. This is not a very realistic representation of actual traffic flow. Also, if the cars are moving at a higher speed, they should maintain longer spacing than when they are moving slower. Thus, the simple CF model does not represent reasonable driving behavior.

3.2.1.3 *CF Model with Spacing as a Function of Speed*

The following vehicle moves in a way that the spacing between the leader–follower is a function of the speed of the follower. When the follower is moving fast, it keeps a larger spacing than when it is going slower. It is expressed as

$$x_n(t) - x_{n+1}(t) = \rho * \dot{x}_{n+1}(t) + d$$

where ρ is the proportionality factor with units of time and d is the distance the vehicle maintains when both are stopped.

This seems a little more reasonable than the simplest CF, but it is still not a good model. Why is it not a good model? This CF model does not consider how far the lead car is away from the follower. Another way to show that it is not a good model is as follows. Take the derivative of this equation

$$x_n - x_{n+1} = \rho * \dot{x}_{n+1} + d$$

$$\dot{x}_n - \dot{x}_{n+1} = \rho * \ddot{x}_{n+1}$$

$$\ddot{x}_{n+1} = \frac{1}{\rho}(\dot{x}_n - \dot{x}_{n+1})$$

This equation says that the acceleration/deceleration of the follower depends only on the difference in speed of the leader–follower pair of vehicles. It does not consider how far the lead car is away from the follower.

An equation in this form $\dot{x}_{n+1} = \frac{1}{\rho}(\dot{x}_n - \dot{x}_{n+1})$ represents the basic form of a stimulus–response equation. The stimulus is the difference in speed of leading and following vehicles, and the response is the acceleration/deceleration of the following vehicle.

The following vehicle cannot immediately respond to a stimulus but would respond after a time lag of T sec. So, the general form of a stimulus–response equation is

$$\dot{x}_{n+1}(t+T) = \frac{1}{\rho}(\dot{x}_n(t) - \dot{x}_{n+1}(t)) = \lambda(\dot{x}_n(t) - \dot{x}_{n+1}(t))$$

where $\lambda = \frac{1}{\rho}$.

If λ = constant, a linear CF model is obtained
If λ = not constant, a nonlinear CF model has resulted.

3.2.1.4 *More Realistic CF Model*

Assuming λ has a constant value is hard to justify. A more reasonable approach would be to relate λ to the speed of the follower and inversely relate it to the spacing between the two vehicles. This would yield a λ value in the following form:

$$\lambda = \alpha \frac{\dot{x}_{n+1}(t)}{[x_n(t) - x_{n+1}(t)]}$$

where α has a constant value, and its value depends on the units used for speed and distance.

3.2.1.5 *Generalizing CF Model*

The realistic model in the previous section was generalized by Gazis *et al.* (1961), in which the speed is raised to a power "*m*" and the spacing to a power of "*l*". This is called the GM5 model in some publications (May, 1990). Thus, λ is directly related to the speed of the follower to the power of "*m*" and inversely related to the spacing between the leader–follower pair to the power of "*l*". This would yield a generalized CF model in the following form:

$$\ddot{x}_{n+1}(t+T) = \alpha \frac{\dot{x}^m_{n+1}(t)}{[x_n(t) - x_{n+1}(t)]^l}\left[(\dot{x}_n(t) - \dot{x}_{n+1}(t)\right]$$

Using different m and l values, one gets different CF models, as shown in Figure 3.2 (May *et al.*, 1967).

3.2.2 *Car Following in Simulation Software*

Car following models discussed above are used to study traffic flow stability. They are rarely used in traffic simulation software. Car following models used in simulation packages not only find the acceleration/deceleration of the following vehicles but also provide additional inputs and constraints to the movement of vehicles. The main constraints are collision avoidance requirements, emergency stop criteria, and physical limitations on vehicle performance (Benekohal *et al.*, 1989; Aycin *et al.*, 1998). Commercial simulation software uses car following models that are unique to them, and often, limited information is available about them.

For detailed derivation of car following models for five non-commercial simulation programs (NETSIM, CARSIM, INTRAS, FRESIM, and INTELSIM) and comparison of their performance vs field data, see Aycin *et al.* (1999).

Vissim (Fellendorf & Vortisch, 2010) is a popular commercial traffic simulation software and primarily uses either the Wiedemann 74 or Wiedmann 99 psycho-physical car following model. The Wiedemann 74 model is typically used for urban and stop-and-go traffic, while the Wiedemann 99 is more suitable for freeway traffic. The models assume that the following driver's speed and distance are influenced by the leading vehicle. The models combine the psychological desire of the following driver to maintain a specific safety distance with the physical ability to perceive changes in speed and distance relative to the preceding vehicle.

Aimsun is another popular commercial traffic simulation software (Casas *et al.*, 2010), and it uses a car following model that is based on the Gipps (1981, 1986) model. Vehicles are updated based on vehicle behavior in car following and lane changing models. Drivers try to reach their desired speed considering the traffic, roadway, and control conditions.

Paramics is another popular commercial software (Skyes, 2010). It uses pyscho-physical car following model, a variant of the Wiedmann model. Its car following model is based on maintaining a safe speed and distance between follower and leader while considering the distance between them, their speed difference, and a minimum headway.

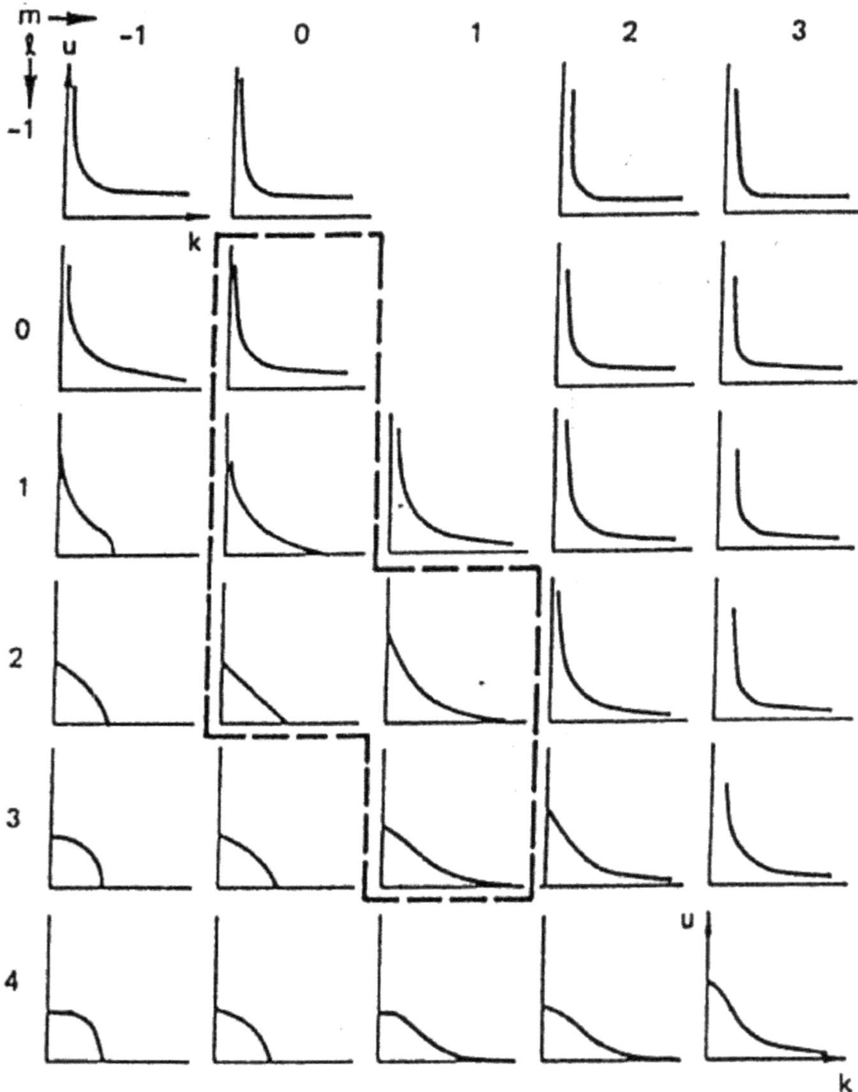

Figure 3.2. Matrix of speed–density relationship for various m, l combinations of the general car following equation (dashed lines enclose limiting values of l and m used).

Source: From May, Jr., A. D., and H. E. M. Keller. Non-Integer Car Following Models. Highway Research Record, No. 199, 1967, Figure 4, p. 24. Copyright, National Academy of Sciences. Reproduced with permission of the Transportation Research Board.

3.3 Relationship between Macroscopic and Microscopic Models

Some of the microscopic models have a corresponding macroscopic model. This means you can derive some of the macroscopic models from a corresponding car following model. For example, you can derive Greenshields and Greenberg models from the corresponding car following model. For example, if $m = 0$ and $l = 2$, you get the Greenshields model. Similarly, if $m = 0$, $l = 1$, you get the Greenberg model as shown in the following.

Deriving the Greenberg model from the corresponding CF model.

In the general car following equation, put 0 for "m" and 1 for "l", to get the following equation:

$$\ddot{x}_{n+1}(t+T) = \alpha \frac{\dot{x}o_{n+1}(t)}{\left[x_n(t) - x_{n+1}(t)\right]^1}\left[\left(\dot{x}_n(t) - \dot{x}_{n+1}(t)\right)\right] = \alpha \frac{\left[\left(\dot{x}_n(t) - \dot{x}_{n+1}(t)\right)\right]}{\left[x_n(t) - x_{n+1}(t)\right]^1}$$

Integration of it would yield

$$\dot{x}_{n+1}(t + T) = \alpha \, \mathrm{Ln}[x_n(t) - x_{n+1}(t)] + C$$

In a steady-state traffic condition, speed (u) at time "t" and speed at time "$t + T$" will be the same. So, we can write

$$\dot{x}_{n+1}(t) = u = \alpha \, \mathrm{Ln}[x_n(t) - x_{n+1}(t)] + C$$

The term $[x_n(t) - x_{n+1}(t)]$ represents the spacing between vehicles, and it is equal to $1/k$.

Thus,

$$u = \alpha \, \mathrm{Ln}(1/k) + C$$

Using boundary conditions (when $u = 0$, $k = k_j$), one can find C:

$$0 = \alpha \ \mathrm{Ln}(1/k) + C$$

$$C = -\ \alpha \ \mathrm{Ln}(1/k_j)$$

Substituting this value for C yields

$$u = \alpha \mathrm{Ln}(1/k) - \alpha \mathrm{Ln}(1/k_j) = \alpha \mathrm{Ln}(k_j / k)$$

This is in the form of the Greenberg model. To find the value of "α ", differentiate q with respect to k

$$q = \alpha * k * \mathrm{Ln}\left(k_j / k\right)$$

$$\frac{dq}{dk} = \alpha * \mathrm{Ln}\left(k_j / k\right) - \alpha$$

Setting this equal to zero results in Ln (k_j/k)=1. Substituting 1 for Ln(k_j/k) in the equation gives the speed at max flow (u_c). Thus, $\alpha = u_c$. Substituting this in the above equation gives the Greenberg model:

$$u = u_c * \mathrm{Ln}\ (k_j/k)$$

3.4 Mesoscopic Approach (Models)

Mesoscopic approach (models) may be a combination of microscopic and macroscopic approaches or a representation of traffic flow in a different format that may resemble macroscopic or microscopic approach. Mesoscopic approach is not as clearly defined as the other two approaches. It might be a group of cars that are treated as one entity and moved as a group. Another example would be in modeling traffic in a large network; at the network level, a macroscopic approach is used (e.g. traffic volumes come from the traffic assignment step in the four-step planning process), but at the intersection level, a microscopic approach is used. Another example of the mesoscopic approach is when a platoon of vehicles is modeled as one entity (one unit) and the unit may have several cars, but the cars are not modeled individually. It is advised to ask what do they mean when they use the phrase mesoscopic approach.

Barcelo (in *Fundamentals of Traffic Simulation*) considers basically two main approaches to mesoscopic simulation: those that pack vehicles in groups (platoon) such as CONTRAM (Leonard *et al.*, 1989) and those that use a simplified dynamics of individual vehicles, such as DYNASMART (Jayakrishnan *et al.*, 1994), DYNAMIT (Ben-Akiva *et al.*, 1997, 2001, 2002), DTASQ (later on Dynameq) (Florian *et al.*, 2001, 2002; Mahut *et al.*, 2003a, 2003b, 2004), and MEZZO (Burghout *et al.*, 2005).

An example of mesoscopic models is Dynameq (which stands for dynamic equilibrium) which is a simulation-based dynamic traffic assignment (DTA) model (Mahut & Florian, Chapter 9 of Fundamentals of Traffic Simulation by Barcelo). The following simplified car following model is used to reduce the computation time:

$$x_{n+1}(t) = \min\left[x_{n+1}(t-\varepsilon) + \varepsilon V, x_n(t-R) - L\right]$$

where $x_{n+1}(t)$ is the position of the following vehicle at time, $x_n(t - R)$ is the position of the lead vehicle at time $(t - R)$, V is the free flow speed, R is the response time of the driver, ε is an arbitrary short time interval, and L is the effective vehicle length. With this logic, the vehicle position is the minimum of the first or the second term. The first term gives the position of the vehicle if it can be moved forward with a speed of V during the time of ε, and the second term is the position the follower can be moved without colliding with its leader. As you can see, this model is very simple and it only defines the position of each vehicle in time rather than the vehicle speed or acceleration. This simplified car following logic assumes that speed is constant over a short time period, and that is in contrast with a traditional car following model that uses a constant acceleration to compute speed and then update the position of a vehicle. From the above equation, one can derive the time a vehicle will reach a given point (Mahut above). The given point may be the entrance or exit to a link. With this logic, you don't need to update the position of the vehicle every second, rather you compute the position when an event happens (the event may be the time a vehicle enters the link or when it leaves the link).

This simplified car following logic is used with a DTA algorithm that assigns the vehicles to a path. A driver selects his optimal path from all available paths that approximately minimizes his travel time. The path input flow is determined by a variant of the method of successive averages (MSA) or gradient-like algorithm, which is applied to each pair or O-D pair. The process is repeated until a given number of iterations is reached or a desired accuracy is achieved.

Exercises

Part 1. Questions

1. Define macroscopic, microscopic, and mesoscopic traffic flow modeling approaches. For each, state what is modeled and provide one example output variable.

2. Explain the hydrodynamic (continuum) analogy in macroscopic traffic modeling. What does it mean to treat traffic as a compressible continuum medium?
3. State the fundamental relationship of traffic flow and interpret each variable. Why is it called a "fundamental" relationship?
4. Write the continuity equation and explain, in words, what conservation of vehicles means for a roadway segment with no entrances/exits.
5. Distinguish between first-order and higher-order macroscopic models. What is the "instantaneous speed adaptation" assumption in first-order models?
6. Explain why higher-order models (e.g. Payne–Whitham) introduced relaxation and anticipation. What driver behavior does each term represent?
7. Briefly describe the wave propagation issue raised by Daganzo regarding some higher-order models and how ARZ-type models address it (conceptually).
8. Explain why Greenshields' speed–density relationship is historically important. List one advantage and two limitations when applied to modern freeways.
9. Compare single-regime, two-regime, three-regime, and four-regime speed–density models in terms of flexibility and practical challenges (e.g. breakpoint selection and discontinuity).
10. Describe the stimulus–response structure of car following (CF) models. Identify what typically serves as "stimulus" and what serves as "response".
11. Differentiate linear vs nonlinear car following models in terms of sensitivity. Give one example factor that can make sensitivity nonlinear.
12. List at least three additional constraints that commercial simulation software must include beyond basic CF equations (e.g. collision avoidance). Why are these important?
13. Explain, conceptually, how some macroscopic models can be derived from microscopic CF models (link between micro and macro).
14. Define mesoscopic modeling and explain why the term can be ambiguous. Provide two different interpretations/examples given in this chapter.
15. Greenshields derivation: Starting from $u = u_f(1 - k/k_j)$ and $q = uk$, derive $q(k)$ and show that it is an upside-down parabola.

Part 2. Problems

Problem 1. Assume the Greenshields model is valid. For a roadway with free-flow speed $u_f = 65$ mph and jam density $k_j = 200$ veh/mi/ln, compute (a) critical density k_c, (b) critical speed u_c, and (c) capacity q_c. Then, compute the corresponding average headway at capacity (sec/veh). Comment on reasonableness.

Problem 2. Using the same u_f and k_j as the previous problem, compute speed u and flow q at density $k = 60$ veh/mi/ln.

Problem 3. Consider the CF model with spacing as a function of follower speed: $x_n - x_{(n+1)} = \rho \cdot \dot{x}_{(n+1)} + d$. (a) Differentiate to obtain an acceleration equation. (b) Identify the stimulus and response terms.

Problem 4. Generalized (GM-type) CF: $\ddot{x}_{(n+1)}(t + T) = \alpha \cdot (\dot{x}_{(n+1)}(t))$ ^ m / $(x_n(t) - x_{(n+1)}(t))$ ^ $\ell \cdot (\dot{x}_n(t) - \dot{x}_{(n+1)}(t))$. For $m = 0$ and $\ell = 1$, show that the sensitivity depends inversely on spacing.

Problem 5. Deriving Greenberg: Starting from the generalized CF model, set $m = 0$ and $\ell = 1$ and outline the key steps that lead to $u = u_c \ln(k_j/k)$.

Problem 6. Greenberg numerical: Let $u_c = 45$ mph and $k_j = 180$ veh/mi/ln. Compute speed u at density $k = 60$ veh/mi/ln.

Problem 7. For freeway traffic condition where free flow speed is 75 mph, capacity is 2400 pc/hr, and jam density is 220 pc/hr. Would a linear speed–density model (Greenshields model) be a good choice? Support why not.

Problem 8. Would the Highway Capacity Manual (HCM) model work for very congested freeway traffic conditions? If not, what model would you use?

Problem 9. This nonlinear relationship between speed (u in mph) and density (k in vpm) was obtained for a freeway traffic $u = 72.6\text{e}{-}^{0.031k}$.

a. Whose model is this?
b. Plot speed–density graph.

c. Find speed–flow relationship and plot it.
d. Find density–flow relationship and plot it.
e. What is free flow speed?
f. What is speed at capacity?
g. What is jam density?
h. What is critical density (density at capacity)?

References

Avrenli, K. A., Benekohal, R. F., & Ramezani, H. (2012). Four-regime speed–flow relationships for work zones with police patrol and automated speed enforcement. *Transportation Research Record: Journal of the Transportation Research Board, 2272*(1), 35–43.

Aw, A. & Rascle, M. (2000). Resurrection of "second order" models of traffic flow. *SIAM Journal on Applied Mathematics, 60*(3), 916–938.

Aycin, M. F. & Benekohal, R. F. (1998). Linear acceleration car-following model development and validation. *Transportation Research Record: Journal of the Transportation Research Board, 1644*(1), 10–19. Washington, DC.

Aycin, M., & Benekohal, R. F. (1999). Comparison of car-following models for simulation. *Transportation Research Record: Journal of the Transportation Research Board, 1678*(1), 116–127. Washington, DC.

Ben-Akiva, M., Bierlaire, M., Bottom, J., Koutsopoulos, H. N., & Mishalani, R. G. (1997). Development of a route guidance generation system for real-time application. *Proceedings of the 8th IFAC Symposium on Transportation Systems, 30*(8), 405–410, Chania, Crete.

Ben-Akiva, M., Bierlaire, M., Burton, D., Koutsopoulos, H. N., & Mishalani, R. G. (2001). Network state estimation and prediction for real-time traffic management. *Networks and Spatial Economics*, 1, 293–318.

Ben-Akiva, M., Bierlaire, M., Koutsopoulos, H. N., & Mishalani, R. (2002). Real time simulation of traffic demand-supply interactions within DynaMIT. In Gendreau, M., Marcotte, P. (eds.), *Transportation and Network Analysis: Current Trends. Applied Optimization*, Vol. 63. Springer, Boston, MA.

Benekohal, R. F. & Treiterer, J. (1989). A car following model for simulation of traffic flow in normal and stop-and-go conditions. TRR 1194, TRB, *National Research Council*, Washington, DC.

Burghout, W., Koutsopoulos, H. N., & Andréasson, I. (2005). Hybrid mesoscopic–microscopic traffic simulation. *Transportation Research Record: Journal of the Transportation Research Board, 1934*(1), 218–225.

Casas, J., Ferrer, J. L., Garcia, D., Perarnau, J., & Torday, A. (2010). Chapter 5: Traffic simulation with Aimsun. In Barcelo, J. (ed.), *Fundamental of Traffic Simulation*. Springer, New York, pp. 173–233.

Daganzo, C. F. (1994). The cell transmission model: A dynamic representation of highway traffic consistent with the hydrodynamic theory. *Transportation Research Part B: Methodological*, *28*(4), 269–287.

Drake, J., Schofer, J., & May, A. D., Jr. (1967). A statistical analysis of speed-density hypotheses. *Proceedings of Third International Symposium on Theory of Traffic Flow* (New York, June 1965), pp. 112–117. Published as Vehicular Traffic Science, American Elsevier, New York.

Edie, L. C. (1961). Car-following and steady-state theory for noncongested traffic. *Operations Research*, *9*(1), 66–76.

Fellendorf, M. & Vortisch, P. (2010). Chapter 2: Microscopic traffic flow simulator VISSIM. In Barcelo, J. (ed.), *Fundamental of Traffic Simulation.* Springer, New York, pp. 63–129.

Florian, M., Mahut, M., & Tremblay, N. (2001). A hybrid optimization-mesoscopic simulation dynamic traffic assignment model. *Proceedings of the 2001 IEEE Intelligent Transportation Systems Conference*, pp. 118–123, Oakland.

Florian, M., Mahut, M., & Tremblay, N. (2002). Application of a simulation-based dynamic traffic assignment model. *European Journal of Operational Research*, *189*(3), 1381–1392.

Gazis, D. C., Herman, R., & Potts, R. B. (1959). Car-following theory of steady state flow. *Operational Research*, *7*(4), 499–505.

Gazis, D. C., Herman, R., & Rothery, R. W. (1961). Nonlinear follow-the-leader models of traffic flow. *Operational Research*, *9*(4), 545–567.

Gerlough, D. L., & Huber, M. J. (1975). Traffic Flow Theory: A Monograph. Transportation Research Record, Special Report 165, TRB, *National Academies*, Washington, DC.

Gipps, P. G. (1981). A behavioural car-following model for computer simulation. *Transportation Research Part B: Methodological*, *15*(2), 105–111.

Gipps, P. G. (1986). A model for the structure of lane-changing decisions. *Transportation Research Part B: Methodological*, *20*(5), 403–414.

Herman, R., Montroll, E. W., Potts, R. B., & Rothery, R. W. (1959). Traffic dynamics: Analysis of stability in car-following. *Operational Research*, *7*(1), 86–106.

Herman, R. & Potts, R. B. (1961). Single-lane traffic theory and experiment. In *Proceedings of Symposium on the Theory of Traffic Flow*. Elsevier, Amsterdam, pp. 120–146.

Herman, R. & Rothery, R. W. (1965). Car following and steady state flow. In *Proceedings of Second International Symposium on the Theory of Road Traffic Flow*, London, 1963. OECD.

Jayakrishnan, R., Mahmassani, H. S., & Hu, Ta-Yin. (1994). An evaluation tool for advanced traffic information and management systems in urban networks. *Transportation Research Part C: Emerging Technologies*, *2*(3), 129–147.

Kometani, E. & Sasaki, T. (1961). Car following theory and stability limit of traffic volume. *Operations Research Society of Japan, 3*(4), 176–190.

Leonard, D. R., Gower, P., & Taylor, N. B. (1989). CONTRAM: Structure of the Model. Research Report 178, Transport and Road Research Lab, Crowthorne, United Kingdom.

Lighthill, M. J. & Whitham, G. B. (1955). On kinematic waves. II. A theory of traffic flow on long crowded roads. *Proceedings of the Royal Society of London. Series A, 229*, 317–345.

Mahut, M., Florian, M., & Tremblay, N. (2003a). Space-time queues and dynamic traffic assignment: A model, algorithm and applications. *Transportation Research Board 82nd Annual Meeting*, Washington, DC.

Mahut, M., Florian, M., & Tremblay, N. (2003b). Traffic simulation and dynamic assignment for off-line applications. *10th World Congress on Intelligent Transportation*, Madrid: ITS World Congress.

Mahut, M., Florian, M., Tremblay, N., Campbell, M., Patman, D., & McDaniel, Z. K. (2004). Calibration and application of a simulation-based dynamic traffic assignment model. *Transportation Research Record: Journal of the Transportation Research Board, 1876*, 101–111. Washington, DC.

May, A. D. (1990). *Traffic Flow Fundamentals*. Prentice Hall, NJ.

May, A. D., & Keller, H. E. M. (1967). Non-integer car following models. Highway Research Record, 199, Transportation Research Board, Washington, DC, pp. 19–32.

Payne, H. J. (1971). Models of freeway traffic and control. In Bekey, G. A. (ed.), *Mathematical Models of Public Systems*, Vol. 1, pp. 51–61. Simulation Council, La Jolla.

Pipes, L. A. (1953). An operational analysis of traffic dynamics. *Journal of Applied Physics, 24*(3), 274–281.

Richards, P. I. (1956). Shock waves on the highway. *Operations Research, 4*(1), pp. 42–51.

Skyes, P. (2010). Chapter 4: Traffic simulation with paramics. In Barcelo, J. (ed.), *Fundamental of Traffic Simulation*. Springer, New York, pp. 131–171.

Treiterer, J. (1967). Improvement of traffic flow and safety by longitudinal control. *Transportation Research, 1*(3), 231–251. Elsevier Pub Co.

Treiterer, J. & Myers, J. A. (1974). The hysteresis phenomena in traffic flow. In *Proceedings of 6th International Symposium on Transportation and Traffic Theory*, Elsevier Pub. Co, New York, pp. 13–38.

Underwood, R. T. (1961). Speed, volume, and density relationships: Quality and theory of traffic flow. *Yale Bureau of Highway Traffic*, 141–188.

Whitham, G. B. (1974). *Linear and Nonlinear Waves*. Wiley & Sons, NY.

Chapter 4

Traffic Simulation Types

Simulation is a tool, like other tools such as mathematics, statistics, and operation research (OR), used to analyze how a complex system performs under different conditions. It usually does not directly optimize the performance of a system, but one can use it repeated times to find a near optimal solution. Simulation should be used when other mentioned tools do not work. When is it appropriate to use a simulation model is summarized in Table 4.1.

4.1 Categories of Traffic Simulation

Traffic simulation models can be categorized in a few ways depending on the criteria used (Barcelo, 2010; Law & Kelton, 2000). Broadly speaking, traffic simulations can be categorized based on how they deal with factors such as stochasticity, scale of modeling, treatment of flow, and advancement of the simulation clock. This categorization is summarized in Table 4.2.

4.1.1 *Model-Based Traffic Flow Granulation*

Simulation models based on the approach used for modeling traffic flow may be categorized as microscopic, macroscopic, and mesoscopic models, as shown in Table 4.3.

In the previous chapter, we discussed how traffic flow may be treated as a continuous flow of fluids in channels. When the flow of traffic is

Table 4.1. When simulation should/should not be used.

Should Be Used	Should Not Be Used
Real-world experimentation is not possible, very expensive, not feasible to repeat several times, or unsafe	If you have other tools (such as math, stat, or OR) available
Need to assess performance of system under many conditions to assess different alternatives or to answer what-if questions	If you don't know how to properly model and simulate your system
System is too complex to express in terms of mathematical equations or there is no analytical approach available	If you don't have adequate training and skill to use it correctly

Table 4.2. Traffic simulation categories.

Traffic Simulation Categories may be based on

1) How traffic flow is granulated
   - Microscopic
   - Macroscopic
   - Mesoscopic

2) How events are modeled
   - Discrete
   - Continuous
   - Combination of the two

3) How simulation clock is advanced
   - Time based
   - Event based
   - Both

4) How variability is handled
   - Deterministic simulation
   - Stochastic simulation
   - Monte Carlo simulation

assumed to resemble the flow of a continuous media, such as water (in general, the continuum flow models), the simulation models using this approach are called macroscopic models. However, when traffic is treated as a collection of many individual units or entities (cars, peds, etc.), the

Table 4.3. Traffic simulation categories based on traffic flow granulation.

Category	What Is Modeled	Example
Microscopic Models	An entity (a unit) such as a vehicle, a driver, a pedestrian, a bike, a train, a bus, or a single user	A driver with known reaction time, desired speed, and car following distance
Macroscopic Models	Traffic stream characteristics such as flow, average speed, or density	Average speed of the traffic stream and traffic volume over certain time period
Mesoscopic Models	Depending on what they mean by phrase. It may be a mix of micro and macro approaches, a simplified version of micro approach, or other actions to balance accuracy with efficiency	A platoon of vehicles is treated as a unit, and these units are modeled instead of individual vehicle

simulation models using this approach are called microscopic approach. So, traffic simulation models may be categorized as

a. microscopic;
b. macroscopic;
c. mesoscopic.

In microscopic approach, the driver–vehicle units are generated with known characteristics. Then, they are moved longitudinally by car following logic, and laterally by lane changing logic. Statistics at the unit level are tallied to create system output.

In general, the macroscopic simulation models are faster than microscopic models, but they provide less detailed information compared to microscopic models.

The approach that is a hybrid of microscopic and macroscopic approaches is called mesoscopic approach. The following approaches are considered mesoscopic (Kessels, 2019): cluster models (Kerner & Konhauser, 1994), gas kinetic continuum models (Prigogine & Herman, 1971), and macroscopic models derived from them (Joueiai *et al.*, 2015). In the cluster models, adjacent vehicles are put in groups that share a specific property (like a platoon of vehicles). The size of a cluster may change, but vehicles inside a cluster have the same speed. Vehicles may join or leave the cluster. The speed and location of clusters are updated, not the speed or location of the individual vehicles in the cluster.

Mesoscopic approach in Aimsun simulation model means having a discrete-event simulation that updates the state of the system when an event (e.g. vehicle arrives or departs or change in traffic light status) occurs and moves the individual vehicles using simplified car following, lane changing, and gap acceptance models (Casas *et al.*, 2010).

Other types of simulation models that may be considered mesoscopic are cell transmission and cellular automata models. In cell transmission model, the roadway is divided into small segments (cells), and each cell has input flow, output flow, number of vehicles in the cell (Daganzo, 1994). Adjacent cells are linked to each other by the amount of traffic they send out to the downstream cell or receive from the upstream cell. The length of the cell is specified by the model developed and can be from a few 100 ft to miles. For example, WorkzoneQ-Pro software uses a cell length of 250 ft for the simulation of traffic on arterial streets and 500 ft on the freeways. WorkZoneQ-Pro updates the speed, flow, and density information of each cell every 2 sec for the arterials and every 4 sec for the freeways (Benekohal *et al.*, 2013).

Cellular automata concept is also used in traffic simulation (Nagel & Schreckenberg, 1992; Rickert *et al.*, 1996; Maerivoet & De Moor, 2005; Wolf, 1999; Blue & Adler, 2001; Los Alamos National Laboratory, 1995; Bham & Benekohal, 2004). Roadway is divided into small sections (equal size), and roadway is a lattice of these cells. The cells are small enough to hold a single vehicle in them. Each cell is either occupied by a vehicle or is empty. The way these cells are occupied or unoccupied allows for the monitoring of the movement of vehicles. There are rules for moving vehicles. If a vehicle is moving slower than the maximum speed, it accelerates. If it is moving faster and closing the gap to the lead car and itself, it decelerates. The vehicle may coast when certain conditions are met.

4.1.2 *Models Based on Handling Events and Processes*

If we assume that traffic flow is composed of individual units (entities), we should use discrete event simulation models (such as CORSIM and VISSIM). However, if we assume the flow of traffic on roads is analogous to the flow of water in channels (continuous media), we should use a continuous simulation model. So, they can be classified into these categories:

a. discrete-event simulation (DS);
b. continuous simulation (CS);
c. combination of DS and CS.

In discrete-event simulation, the system evolves over time by being updated at certain points in time. The system can change at only a countable number of points in time and when an event happens (such as the arrival or departure of a vehicle at an intersection) (Law & Kelton, 2000). Some events such as the arrival of a vehicle changes the state of the system instantaneously and some events such as the end of simulation may not change the state of the system.

In continuous simulation, the time is advancing continuously and changing the state of the variable (system). Typically, continuous simulation models involve differential equations to express the rate of change in state variables with time (Law & Kelton, 2000).

To model complex real-world conditions, it may be advantageous to use it as a combination of DS and CS. An example of such a system is given by Pritsker (1995) as an oil refinery where the fuel moves in pipes for a while and then is distributed by tankers. Some simulation languages such as SLAM (Pritsker, 1995) can model the combined discrete-continuous systems.

4.1.3 *Models Based on Advancing Simulation Clock*

Simulations may be grouped based on how the system clock is advanced. Advancing the system clock implies that an input, output, or event in the system occurred that may change the system status. Time advancing may be based on one of the following mechanisms:

a. time-based;
b. event-based;
c. combination of time and event.

One way of a time-advancing mechanism is updating the system variables in regular time intervals (e.g. every 1 sec) and thus updating the system status at the end of that interval. In this case, the system does not change during the interval. In the event-based time advancing mechanism, the system status is updated when an event such as the arrival or departure of a vehicle, the end of a process, or reached the end of the simulation occurs. A third way of advancing the system clock is to do a combination of time-based and event-based. In this case, the system status is updated at the end of each time interval in addition to the times an event occurred. If no event happens in the middle of the interval, the system will be updated at the end of the interval.

4.1.4 *Models Based on Handling Variability*

Simulation models may be categorized based on how they handle the variability, if exist, in the simulation. The categories are

a. deterministic simulation;
b. stochastic simulation;
c. static and Monte Carlo simulation.

4.1.4.1 *Deterministic Simulation*

A deterministic simulation assumes that the behaviors of all entities are predefined, and they have known values. For example, the driver's reaction time may be a known constant value; or it may change from driver to driver based on pre-assigned values (it is not randomly assigned to the driver). In deterministic simulation, the input values are determined and remain constant at least for a known period. The input values and their effects on the system can easily be traced. The outcome is constant for the given input. There are no stochastic variables in deterministic simulation that change the outcome. This type of simulation is not of much interest when there is stochasticity in the system. For example, in deterministic simulation, one may assume that vehicle characteristics and driver desired speed, headway, reaction time, desired following distance, acceleration, and deceleration are constant. These are not very reasonable assumptions if the simulation model is expected to realistically represent real-world traffic conditions.

4.1.4.2 *Stochastic Simulation*

In stochastic simulation, the values assigned to an entity are not constant but come from a distribution that represents that variable. As input changes to represent stochasticity in that variable, the output may also be affected by the change. In stochastic simulation, the behavior of a driver may have a range, or a probability distribution, and its value is randomly assigned to drivers. For example, the reaction time of drivers comes from a distribution are randomly assigned to them. Similarly, driver characteristics such as desired speed, headway, reaction time, desired following distance, acceleration, and deceleration vary from driver to driver. These variations are essential in realistically representing the actual traffic conditions in the simulation model. Since the input to a stochastic simulation has stochastic elements, the simulation experiment must be run properly to deal with the variability (refer to the batch means vs replication discussion). This is a very important point that sometimes is ignored or not well understood.

4.1.4.3 *Static and Monte Carlo Simulation*

A static simulation model represents the system at a particular time or a system in which time plays no role in its outcome (Law & Kelton, 2000). Monte Carlo simulations are generally static rather than dynamic (Law & Kelton, 2000). Monte Carlo simulation may use random variates to deal with certain stochastic or deterministic problems, but passage of time plays no substantive role (Law & Kelton, 2000). Thus, Monte Carlo simulation is neither deterministic nor stochastic but has some elements from each. It considers the distribution of input variables and finds the output corresponding to repeated random samples of input variables. When a system has input variables that are random, Monte Carlo simulation is often superior to a deterministic simulation (Sokoloski & Banks, 2010). For Monte Carlo simulation (or Monte Carlo method), a four-step process is suggested by Sokolowski:

Step 1. For each input random variable, define a distribution.
Step 2. From the distributions, generate inputs randomly.
Step 3. For that set of inputs, compute the outcome.
Step 4. Combine the outcomes from the individual computations.

In Step 2 from the distribution of each random variable, many random samplings will be done to develop a vector of inputs for that variable. The number of samples will depend on the desired accuracy of the outcome. Refer to determining sample size elsewhere. It may seem that you are doing sensitivity analysis on the effects of changing randomly selected values of a variable. There may be several of these variables, and each variable may have many different randomly selected values. So, you need to run your simulation for combinations of these variables to determine the outcome for the combination of variables.

Exercises

1. Explain why simulation is considered a tool similar to mathematics, statistics, and operations research.
2. Why does simulation generally not directly optimize system performance?
3. List three situations where simulation should be used and three where it should not be used.
4. Define traffic flow granulation in simulation modeling.
5. Define and distinguish between microscopic, macroscopic, and mesoscopic simulation models.
6. Why are macroscopic models typically faster than microscopic models?
7. Compare microscopic and macroscopic simulations in terms of detail, computational cost, data requirements, and applications.
8. Which modeling approach is most appropriate for studying lane changing and gap acceptance? Justify your answer.
9. Which modeling approach is most suitable for regional congestion planning? Why?
10. Describe the structure and purpose of cell transmission models.
11. Explain how cellular automata models represent vehicle movement.
12. Define discrete-event simulation, continuous simulation, and hybrid simulation.
13. Provide two examples of events that change the system state in the traffic simulation.
14. Why are differential equations associated with continuous simulation?
15. Provide an example where a hybrid discrete-continuous model would be appropriate.
16. Explain time-based, event-based, and combined clock advancement mechanisms.
17. What are the advantages and disadvantages of fixed time-step updating?

18. When is combined time-event advancement preferred?
19. Define deterministic simulation and give two unrealistic assumptions if applied to traffic.
20. Define stochastic simulation and explain why multiple replications are required.
21. Compare deterministic and stochastic simulations in realism and computational needs.
22. Define static simulation and explain how Monte Carlo simulation fits this category.
23. Explain why insufficient replications in stochastic simulation may lead to misleading results.
24. Compare Monte Carlo simulation and sensitivity analysis.
25. What are the similarities and differences between deterministic and Monte Carlo simulations?

References

Barceló, J. (ed.) (2010). *Fundamentals of Traffic Simulation.* Springer, New York.

Benekohal, R. F., Ramezani, H., & Avrenli, K. A. (2013). WorkZoneQ User Guide for Two-Lane Freeway Work Zones. FHWA-ICT-13-019, 2013. ideals.illinois.edu.

Bham, G. H. & Benekohal, R. F. (2004) A high fidelity traffic simulation model based on cellular automata and car-following concepts. *Transportation Research Part C: Emerging Technologies, 12*(1), 1–32.

Blue, V. J. & Adler, J. L. (2001). Cellular automata microsimulation for modeling bi-directional pedestrian walkways. *Transportation Research Part B, 35*(3), 293–312.

Casas, J., *et al.* (2010). Chapter 5: Traffic simulation with Aimsun. In Barcelo, J. (ed.), *Fundamental of Traffic Simulation.* Springer, New York, pp. 173–233.

Daganzo, C. F. (1994). The cell transmission model: A dynamic representation of highway traffic consistent with the hydrodynamic theory. *Transportation Research Part B: Methodological, 28*(4), 269–287. https://doi.org/10.1016/0191-2615(94)90002-7.

Joueiai, M., Leclercq, L., van Lint, J. W. C., & Hoogendoorn, S. P. (2015). A multi-scale traffic flow model based on the mesoscopic LWR model. *Transportation Research Record, Journal of the Transportation Research Board, 2491*, 98–106.

Kerner, B. S. & Konhauser, P. (1994). Structure and parameters of clusters in traffic flow. *Physical Review E*, *50*, 54. https://doi.org/10.1103/PhysRevE.50.54.

Kessels, F. (2019). Chapter: Mesoscopic models. In *Traffic Flow Modelling: Introduction to Traffic Flow Theory through a Genealogy of Models*. Springer International Publishing, part of Springer Nature, Switzerland.

Law, A. M. & Kelton, W. D. (2000). *Simulation Modeling and Analysis* (3rd ed.). McGraw-Hill Inc., New York.

Los Alamos National Laboratory. (1995). *TRANSIMS (Transportation Analysis and Simulation System)*. Los Alamos National Laboratory, Los Alamos, New Mexico.

Maerivoet, S. & De Moor, B. (2005). Cellular automata models of road traffic. *Physics Reports*, *419*(1), 1–64. https://doi.org/10.1016/j.physrep.2005.08.005.

Nagel, K. & Schreckenberg, M. (1992). A cellular automaton model for freeway traffic. *Journal de Physique I*, *2*(12), 2221–2229.

Prigogine, I. & Herman, R. (1971). *Kinetic Theory of Vehicular Traffic*. American Elsevier, New York.

Pritsker, A. A. B. (1995). *Introduction to Simulation and SLAM II*. Wiley; Systems Publishing Co., New York, NY.

Rickert, M., Nagel, K., Schreckenberg, M., & Latour, A. (1996). Two lane traffic simulations using cellular automata. *Physica A: Statistical Mechanics and Its Applications*, *231*(4), 534–550. https://doi.org/10.1016/0378-4371(95)00442-4.

Sokolowski, J. A. & Banks, C. M. (2010). *Modeling and Simulation Fundamentals: Theoretical Underpinnings and Practical Domains*. Wiley and Sons, New Jersey.

Wolf, D. E. (1999). Cellular automata for traffic simulations. *Physica A: Statistical Mechanics and Its Applications*, *263*(1–4), 438–451. https://doi.org/10.1016/S0378-4371(98)00536-6.

Chapter 5

What to Do before Simulation

Before simulating a system, explore if there are analytical models or other appropriate ways to assess the system's performance. Also, consider how much you know about the system and its components to be simulated. In general, simulation would not be an appropriate tool when the internal working of the system is not well understood or well defined. Questions listed in the following may help make informed decision before using simulation:

- What are the study goals and system performance measures?
- Are there analytical models adequate for system evaluation?
- Is there a good understanding of the system's functionality?
- Can simulation represent the system and is it the right tool?
- Should I use default values in a simulation package or calibrate them?
- Is the data suitable for calibration/validation of the model and its parameters?

5.1 Inadequacy of Analytical Models and Justification for Simulation

If there are reliable and accurate analytical models, consider using them to analyze the system performance. However, if the analytical models are too simplistic and do not realistically represent your system, then consider using simulation. To use simulation appropriately, one needs to understand the way the real system works. A person who does not understand how the real-world system works may not be able to simulate it properly and may obtain misleading conclusions about the system's performance. Once the system functionality is well understood and its operation is properly characterized, then accurately presenting the system in the simulation environment is crucial. To present the system properly in the simulated environment, one needs to know the input parameters that characterize the system, the components of the system, and how the components interact with each other. Without a clear understanding of the real system's functionality, one cannot represent it properly in the simulation environment.

5.2 Steps to Be Taken before Simulating a System

5.2.1 *Goals and Objectives of Simulating*

The goals and objectives of the simulation study should be clearly defined. The goal might be reducing congestion, evaluating a new roadway design, improving traffic safety, etc. The objective should be clearly stated and the criteria for measuring the performance be decided.

5.2.2 *System and Its Components*

System functionality, key components of the system, and interaction among the components should be known and understood by the modeler. For example, if a highway corridor is simulated, understanding how traffic on the freeway and surrounding streets interact with each other is needed. Components of this corridor might be the freeway itself, ramps and weaving sections, surrounding streets network, intersections, and the traffic control plan used. The extent of the corridor in terms of time and space needs to be defined. How the components affect the whole system.

5.2.3 *Data Collection and Analysis*

Once the system and its boundaries are defined, proper data and adequate data need to be collected and analyzed. Traffic data (such as traffic volume, traffic mix, and origin-destination), network geometry data (such as length and width of road, type of traffic signal, signal timing, and traffic signs), and users' information (such as type of users, drivers or pedestrians, their characteristics, and their car following behavior) should be collected and analyzed.

5.2.3.1 *Data from Field or Reliable Sources*

There should be enough field data, or data from reliable sources, to use for calibration and validation of the simulation model. For example, to simulate a signalized intersection operation, one needs to have data on intersection geometry (e.g. number of thru lanes, turn lanes and their length, width of the road, and grade of the approaches) and traffic characteristics (e.g. vehicular volume on each lane, percent of trucks and buses, no of pedestrians, and no of bikes), traffic signal information (e.g. signal type, detection type, signal indications used, cycle length, green times, phase plan, and offsets), and constrains due to any traffic operation requirements (e.g. min green time and left turn movement restrictions). If reliable field data are not available, one should try to find reliable benchmark data that can be used in lieu of the field data. This benchmark data may come from an accurate analytical model or another reliable computer-generated data. For example, a well-established computer-generated data that is used by the profession in assessing systems performance such as a reliable and established simulation model whose performance is widely accepted by the profession.

5.2.3.2 *Data Quality and Quantity*

Data quality and data coverage are very important in assessing the performance of a simulation model. Knowing to what extent data are needed for calibration and subsequently for model validation would help to see if you have a good enough sample size. For example, if you collected 410 speed data from the field and you want to use 240 of them for calibration and the remaining 170 for validation, would these sample sizes provide enough observation (adequate sample size) for your simulation study? If you don't have enough sample size, calibration and validation efforts may not yield a valid model and may lead to misleading results and conclusions.

Having a large set of data with many influential variables or when the effects of each variable cannot be isolated does not mean your sample size is enough. In this case, the quantity may be large enough, but the quality is lacking. For example, if you have many speed and volume data points that come from very similar volume levels, this does not tell you how volume variation affects speed because your data have very limited volume variation. It does not matter how large your data set is; it does not have the volume range you need to study volume–speed relationship.

Similarly, if you have a large data set on speed and volume but it comes from different roadway geometry, traffic conditions, weather conditions, traffic operating conditions, and traffic compositions and the data cannot be divided based on the above-mentioned variables, this data set is large but not very useful to study speed–flow relationship. It represents a mixture of all factors that are not very helpful in characterizing the effects of each of those conditions.

5.2.3.3 *Data Recency*

It is important to have up-to-date data to study the current and future conditions. The data should be recent enough to represent the current or anticipated future condition. Traffic conditions and vehicle characteristic and the infusion of technology, vehicle design, roadway design changes, and the behavior of drivers may change over time. The data collected 20 years ago may not represent the current vehicle fleet, driving population, roadway conditions, and vehicle characteristics very well. For example, studying traffic flow characteristics based on the data that was collected when all lanes had to go through manual toll plaza gates may not be a good data set for toll roads that have bypass lanes where vehicles do not stop.

5.2.4 *Modeling Approach and Software Selection*

The modeling approach (microscopic, macroscopic, and mesoscopic) should be decided to achieve the stated goals and objectives. If a simulation package (software) is to be selected, its fidelity, input variables, output parameters, ease of use, costs, training, and capabilities and limitations need to be considered.

5.2.5 *Conceptual Model Building*

Conceptual model attempts to represent the actual model on a computer screen or a piece of paper using flow charts or process diagrams and may include event lists, logical rules, and constraints of the systems. For example, the conceptual model for a traffic network could be a sketch of a network on a piece of paper that shows the nodes and links and entry and exit points to the network.

5.2.6 *Calibration and Validation*

First, the parameters to be calibrated should be identified (such as speed, delay, travel time, headways, car following, and lane changes). Then, the data necessary and the approach for calibration of them, as well as the threshold for acceptable differences, need to be decided. Due to lack of proper data, one may end up using the default values of the simulation package. Then, the validation approach and parameters used, and the sample size need to be decided. More on these topics is discussed in the validation chapter.

5.2.6.1 *Calibration vs Using Default Values*

Is there someone in your staff who knows how to calibrate and validate the simulation models? If it is done by outside consultants, is there someone with enough knowledge to specify how to simulate appropriately the conditions you are studying? Should there be any calibration for local conditions, or should the default values be used? Would default values be appropriate for your local conditions? If calibration is needed, what variables should be calibrated?

When you don't have reliable local data, not calibrating should be considered as a viable option. The default values represent the average conditions that the modeler had in mind, and using them would be better than using inappropriate data for calibration.

5.2.6.2 *Choosing Calibration Parameters*

This task is very critical, and many factors play a role in choosing the calibration parameters. One of the factors to consider is how critical this parameter is in the simulation model for your condition. For example, if you are dealing with freeway models, the start-up time and lost time that are important at intersection modeling may not be as important for freeway traffic. Another factor is how sensitive the outcomes are to this parameter. If a parameter is not that critical for your project, you may skip that and choose the ones that are more important. Another factor is the availability of field data or benchmark data. If the field data are not available, you cannot calibrate that parameter.

5.2.7 *Design Experiments*

Experimental design is an important step, and if done properly, one can run the necessary information from the simulation runs to assess how successful the experiment was. What scenarios should be tested, and what parameters should be changed, and what outputs should be collected? The number of simulation runs, warm-up time, random seeds used, the output collected, etc. need to be determined.

5.2.8 *Output Analysis*

To get an objective evaluation, the output data collected from the simulation runs need to be analyzed carefully using statistical analysis techniques. Subjective evaluations may not yield accurate results. The data analyst should be using proper analysis tools and be familiar with statistical analysis tools.

5.3 Suggestions and Example

Seila *et al.* (2003) provided an example of applying simulation to model operation of an emergency department and discussed the considerations and issues encountered in a real-world project. For more on experimental design and data analysis, one may refer to the books such as Box *et al.* (2005), Kirk (2013), Law and Kelton (2000), Montgomery (2019), Seila *et al.* (2003), and Snedecore and Cochran (1989).

Check List of 10 Things to Do Before Simulation

1. Analytical models should be evaluated first because they are simpler, faster, and less costly.
2. Without understanding internal workings, simulation may misrepresent interactions and produce misleading conclusions.
3. Key questions include study goals, performance measures, adequacy of analytical models, understanding system functionality, appropriateness of simulation, need for calibration, and data availability.
4. Quantity vs quality: Large datasets collected under identical conditions lack variability; mixed-condition data without classification reduces usefulness.

5. Data recency ensures representation of the current vehicle fleet, technology, roadway design, and driving behavior.
6. Clearly defining goals and boundaries prevents scope errors and ensures relevant outputs are collected.
7. Risks of default values include misrepresentation of local driver behavior and system conditions. However, defaults may be preferable when reliable local data is unavailable.
8. Lack of volume variation prevents accurate estimation of speed-flow relationships.
9. Inadequate sample size increases statistical uncertainty and reduces validation credibility.
10. Poor boundary definition may cause unrealistic inflows/outflows and inaccurate queue spillback.

Exercises

1. Why should analytical models be evaluated before deciding to use simulation? Provide two situations where analytical models are preferable.
2. Explain why simulation is inappropriate when the internal workings of a system are not well understood.
3. List and explain the key decision questions that should be asked before conducting a simulation study.
4. Distinguish between data quantity and data quality. Provide two examples where quantity is large, but quality is insufficient.
5. Why is data recency important in traffic simulation? Provide two modern examples.
6. Explain the importance of clearly defining study goals, performance measures, and system boundaries.
7. What risks arise from using default parameter values in simulation software?
8. When might using default values be preferable to calibration?
9. How do you decide what parameters should be calibrated?
10. When financial resources for calibration are limited, how would you decide which parameter should be calibrated?
11. Explain how an inadequate sample size affects calibration, validation, and statistical reliability.
12. Identify five parameters to calibrate for a freeway model and justify each.

13. Why is experimental design critical before simulation? Provide three consequences of poor design.
14. Define warm-up period and explain its importance.
15. Why should output analysis rely on statistical techniques rather than subjective judgment?
16. What statistical tools would you use to compare two design alternatives?
17. You collected 410 speed observations (240 for calibration, 170 for validation). Discuss whether this may be adequate and what additional information is needed.

References

Box, G. E. P., Hunter, J. S., & Hunter, W. G. (2005). *Statistics for Experimenters: Design, Innovation, and Discovery* (2nd ed.). Wiley-Interscience, Wiley, New York, NY.

Kirk, R. E. (2013). *Experimental Design: Procedures for the Behavioral Sciences* (4th ed.). SAGE Publications, Inc., Thousand Oaks, CA.

Law, A. M. & Kelton, W. D. (2000). *Simulation Modeling and Analysis* (3rd ed.). McGraw-Hill Inc., New York.

Montgomery, D. C. (2019). *Design and Analysis of Experiments* (10th ed.). Wiley, New York, NY.

Seila, A. F., Ceric, V., & Tadikamalla, P. R. (2003). *Applied Simulation Modeling*. Thomson, Brooks/Cole, Belmont, CA.

Snedecor, G. W. & Cochran, W. G. (1989). *Statistical Methods* (8th ed.). Iowa State University Press, Ames.

Chapter 6

Calibration Verification and Validation

Model calibration, verification, and validation (CVV) are three important stages in traffic simulation. Calibration is the process of adjusting the parameters of a traffic simulation model to accurately represent the real-world traffic conditions (Law & Kelton, 2000; Banks *et al.*, 2001). Verification is to check if the model behaves as the experimenter assumes it would (Banks *et al.*, 2001; Sokolowski & Banks, 2010). Validation is to test whether the simulation model reasonably approximates a real system (Fishman, 1967; Fishman & Kiviat, 1968; Law & Kelton, 2000; Banks *et al.*, 2001; Sokolowski & Banks, 2010). Figure 6.1 schematically shows these three stages.

Verification also includes checking that the simulation model correctly represents network geometry, control plans, management strategies, and traffic demand (Casa *et al.*, 2010). According to Petty (2010), verification typically answers the following questions:

1. "Does the program code of the executable model correctly implement the conceptual model?
2. Does the conceptual model satisfy the intended uses of the model?
3. Does the executable model produce results when it is needed and in the required format(s)?"

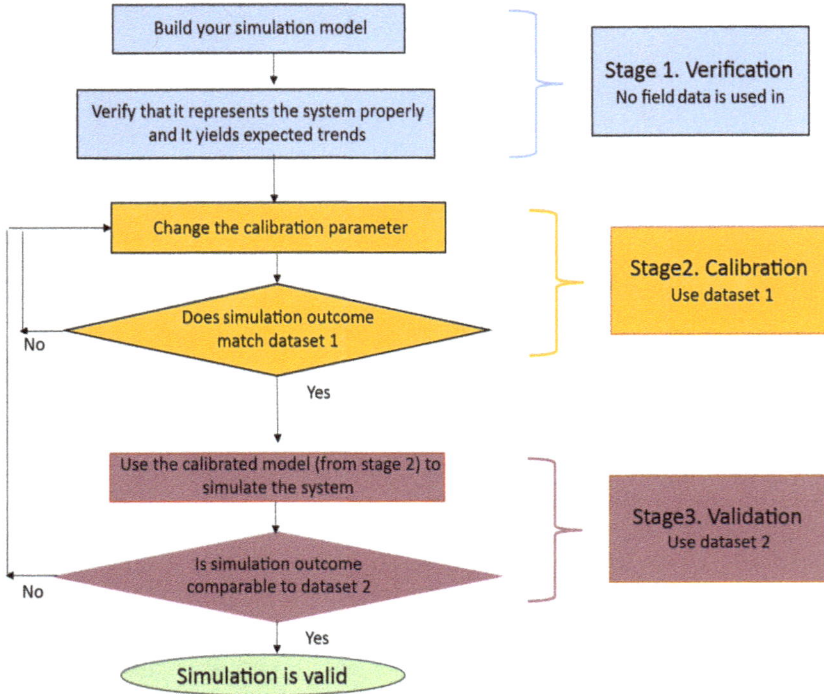

Figure 6.1. Three stages of calibration, verification, and validation of simulation models.

Similarly, validation typically answers the following questions (Petty, 2010):

1. "Is the conceptual model a correct representation of the simuland (a simuland is defined as the real-world item of interest to be simulated)?
2. How close are the results produced by the executable model to the behavior of the simuland?
3. Under what range of inputs are the model's results credible and useful?"

Verification and validation are conceptually different tasks, and initial verification is done before validation. In some cases, verification and validation may be conducted simultaneously (Banks *et al.*, 2001). Validation should not be treated as either/or proposition (Banks *et al.*, 2001) because no model can represent the system or fit the field data perfectly. Some discrepancy is expected, and it should be tolerated.

In addition to verification and validation, an accreditation step was suggested by Petty (2010). Accreditation is not the same as calibration. Accreditation is a non-technical decision process to certify that the model is acceptable for use for a specific purpose. Typical questions to be addressed during accreditation include the following:

1. "Are the capabilities of the model and requirements of the planned application consistent?
2. Do the verification and validation results show that the model will produce usefully accurate results if used for the planned application?
3. What are the consequences if an insufficiently accurate model is used for the planned application?"

6.1 Background on Calibration Verification and Validation

Guidelines developed for calibration of traffic microsimulation models suggest dividing adjustable parameters into those that affect capacity and those that affect route choice (Dowling *et al.*, 2004). They suggested that the capacity parameters should be calibrated first using volume counts at critical bottlenecks. A target for how close the simulation results should be to the field data needs to be specified (an example of such information is the Wisconsin DOT tables for freeway simulation that specifies how close the volume or travel time should be to field data and in what percentage of cases). For route choice calibration, they suggested comparing volume counts from field data to those produced by the simulation model. The parameters that may affect the driver's choice of that route may be travel time, delay, and the cost of the alternate route. After the calibration of capacity and route choice, the system performance calibration is suggested by comparing the overall traffic performance of the model to field data that might be travel time, queue length, and duration of queue. Fine-tuning of link-level parameters may be performed at this stage.

An extensive discussion on the principles of calibration of microsimulation models was presented by Hollander & Liu (2008). They found that many calibration methodologies were not rigorous enough, and most authors mainly used mean values of various traffic measures and did not pay enough attention to the fact that stochasticity is an inherent part of microsimulation models.

Rakha and Wang (2009) developed a procedure for calibrating the Gipps car following model. First, it calibrates the steady state car following model using loop detector data to estimate four traffic stream parameters (free-flow speed, speed-at-capacity, capacity, and jam density). Then, the vehicle acceleration component is calibrated using vehicle specification data.

A three-stage systematic calibration methodology was presented by Hourdakis *et al.* (2003), where in stage one volume is used, in stage two speed, and in stage three effects of ramp metering were assessed. They indicated that the goodness-of-fit tests do not provide sufficient information to identify weaknesses during calibration. They used volume and speed data from the freeway in the calibration and validation of the model. They also compared a calibration process that is significantly automated to the manual one and concluded that they yielded comparable results.

Using unverified models and lack of understanding of the system were among the most frequent causes of simulation failure (Annino & Russell, 1979). Model validation should consist of conceptual validation, computerized validation, operational validation, and the use of adequate and correct data (Sargent, 1982).

For the conceptual validation, the theories, the assumptions, and the relationships used are checked to ensure they are correct and proper for each submodel and for the overall model. Sargent (1982) suggested using the tracing and face validation techniques. In tracing, the behavior of different entities (e.g. vehicles) is traced through each submodel and overall model to determine if the model's logic is correct and the necessary accuracy is obtained. In face validation, the experts in the subjects are asked to evaluate the logic of the submodel and the model and the input–output relationships.

In computerized model validation, it is checked to ensure that the conceptual model is implemented and the computer program runs properly (free of bugs). Each submodel is tested to see if it works properly, and the overall model is executed under different conditions to investigate input and output relations.

Operational validity ensures that the simulation model is a reasonable and accurate representation of the real system with certain levels of confidence. Here, the validation can be done subjectively, such as graphical representation and examination of it, or can be done objectively, such as using statistical techniques.

Use of adequate and correct data could determine the outcome of the CVV efforts. In practice, often there is not enough data available, not enough resources to collect the needed data, or the available data are not of an ideal kind. The user must make the best utilization of the available data.

Verification and validation of microscopic traffic simulation models should be done at two levels: microscopic and macroscopic (Benekohal, 1986, 1991). For verification, the effects of different disturbances on vehicles speed and position should be analyzed. For validation at the microscopic level, the speed change patterns and trajectory plots obtained from simulation models should be compared with those from field data. For validation at the macroscopic level, the average speed, density, and volume from the simulation are compared with those of field data in normal and disturbed traffic flow conditions.

Verification and validation are considered one of the most important and difficult tasks in developing a simulation model (Banks *et al.*, 2001). According to them, "Verification is concerned with building the model right", and "validation is concerned with building the right model". They made several common-sense suggestions for verification. For validation, they suggested that validation is not an either-or proposition because no model can perfectly represent the system under study.

Balci (1998) grouped over 90 different verification and validation (V&V) methods into four broad categories: informal, static, dynamic, and formal (Petty, 2010). Informal V&V methods are more qualitative than quantitative and rely on subjective evaluation by humans. Informal methods include inspection, face validation, Turing test, desk checking, and walkthroughs. Inspection is done in the verification stage, where a group or an individual examines the documents, algorithms, equations used, and the programming language code to make sure they are done appropriately. In face validation, a subject matter expert assesses the behavior of the simulation models and compares them to their knowledge or expected results. The Turing test is an informal validation method where human-generated behavior is compared to system-generated behavior. Desk checking involves a manual review of the model's logic and code to identify potential issues early on. Walkthroughs are a more collaborative approach where the developer presents the model to a group of peers for feedback and error detection.

Static V&V methods analyze the model's structure and the logical consistency of its components without running the simulation. These methods are often applied during the verification phase to ensure that the model is implemented correctly and aligns with the conceptual model. Static methods include data analysis, cause–effect graphing, interface analysis, and traceability assessment. Here, data analysis is a verification method to ensure that the proper data type, data ranges, data structures, and data flow are used.

Dynamic V&V methods evaluate how accurate the simulation results are to benchmark data by numerical comparison. The methods are objective and quantitative. Dynamic methods include sensitivity analysis, predictive validation, comparison testing (statistical methods), graphical comparison, and assertion checking. Sensitivity analysis compares the magnitude and variability of the simulation results to the magnitude and variability of the model (benchmark). Predictive validation compares the outcomes of a simulation to those in benchmark data.

Formal V&V methods involve using mathematically rigorous techniques to prove that the simulation model correctly characterizes the real-world system and accurately represents it. These methods attempt to provide a higher level of confidence in the model's reliability compared to less formal methods. However, formal methods are less practical because the complexity of most modeled systems is too great for the methods to deal with (Balci, 1998). Formal methods included in this category are inductive assertion, predictive calculus, induction, and proof of correctness (Balci, 1998).

Brockfeld *et al.* (2004) compared the performance of 10 different microscopic models. The models were calibrated independently using data from a test track in Japan. The data of the leading car are fed into the model to compute the headway of the following car. They compared the outcome of the models to the field data. During calibration, the error rates ranged from 9% to 24%, and no model was significantly better than other models. For validation, the average error rates were 15.1–16.2%. They concluded that models with more parameters did not necessarily reproduce the field data better. In a special case, the validation error rate was as high as 60% because of an "overfitting" issue in the calibration stage. Overfitting happens when a model is calibrated to represent a particular situation so well that it cannot be generalized to other situations.

Jehn & Turochy (2019) suggested, for the calibration of Vissim for rural freeway work zones, modifying two car following parameters (CC0 and CC1) and using much lower truck acceleration rates than the default (they suggested using 2–3 ft/sec/sec instead of 8, which is the default). They used average speed and mean queue discharge rates for validation of the model. Kan *et al.* (2014) suggested a procedure for calibration of Vissim for freeway work zones where the capacity varies during the analysis period.

Zhao *et al.* (2022) developed a guideline for calibrating and validating a microsimulation model for a freeway to replicate the performance of the automatic queue detection (AQD) system in intelligent work zones (IWZ). For calibration, they suggested the desired speed distribution, car following parameters (CC0, CC1, and CC2), and truck acceleration characteristics in Vissim. A CC0 value of 10 and CC2 of 20.31 was used based on previous studies. Then, CC1 was adjusted to reflect the field data. For validation, traffic volumes were used as well as the warning pattern that the model generated from a simulated AQD system.

Toledo *et al*. (2003) used data from Stockholm under congested traffic conditions to calibrate and validate the traffic simulation tool MITSIMLab. Driving behavior parameters (acceleration, lane changing, and intersection models) and travel behavior components (origin–destination flows and route choice model) were calibrated using speed and flow data. The calibrated model was validated by comparing traffic flows, point-to-point travel times, and queue lengths.

6.2 Model Calibration

Calibration of traffic simulation models is needed to ensure that the model's outcome accurately represents the real-world conditions. This is done by adjusting the parameters of a traffic simulation model. Calibration of microscopic models is different than macroscopic models, as explained in the following.

6.2.1 *Calibration of Microscopic Models*

Microscopic traffic simulation models have unique characteristics because of the interaction among the drivers, vehicles, and roadway environment. The effects of the interactions on traffic flow should be considered in calibration, verification, and validation of the models. A traffic simulation model may not provide accurate results if these three tasks are not properly performed. Sometimes, the users of traffic simulation models calibrate them and do not examine how well the verification and validation steps are carried out. Consequently, they rely on the model performance, assuming that it would represent their condition without calibration and assume that enough verification and validation have been done. Reasons for doing this may be restricted resources and the unavailability of information on CVV. Even when the information is available, for most users, it is difficult to compare the CVV of one model directly to another model because they use different approaches. A systematic approach for calibration, verification, and validation of traffic simulation models should be taken to increase the consistency and reliability of the models.

What parameters should be calibrated for microscopic models? It is not easy to list them because it depends on the microscopic model under consideration. However, there are certain elements that are common to microscopic models, and some of them should be calibrated. It is not feasible to calibrate all the parameters, though it is ideal to do so. Some of the common elements are parameters that affect car following and lane changing behavior, desired speed, reaction time, driving aggressiveness, acceleration/deceleration characteristics, etc. In car following, the headway and spacing between vehicles should be calibrated. In lane changing, the time or distance over which a lane change takes place, the space gaps considered acceptable, and the rate of longitudinal and lateral movement may be calibrated. The desired speed of the user, the reaction time distribution, and the aggressiveness/passiveness of the user may need calibration.

6.2.2 *Calibration of Macroscopic Models*

At the macroscopic level, the goal of calibration is to get model outcomes that are reasonably close to the results one gets from the four-step process in transportation planning. The four steps are trip generation, trip distribution, route choice, and mode choice. Often, it is the origin–destination outcome (results from the first two steps) that is used in calibration. A primer on dynamic traffic assignment (DTA) and its validation is given by Chiu *et al.* (2011). An application of calibration and validation for simulation-based DTA is given by Mahut and Florian (2010) and applied to the Dynameq model. He suggested dividing the process of DTA calibration into two analysis stages: qualitative and quantitative. Qualitative analysis comes before the quantitative analysis. In the qualitative analysis, the quality of the solution and the results should be stable and should make sense. For example, he suggested looking at how the outcomes converge as the number of iterations increased (as expected in a good model).

In the quantitative analysis stages, the results from the simulation model that is producing stable results (the model has reached equilibrium) are compared to the empirical observations. He suggested three categories for sources mismatch between simulation and observed data. One category is the supply side, where parameters related to network characteristics or traffic signal timing need to be examined. The second category is related to routing, where the assignment of roads may not properly reflect the driver's behavior. And the third category is issues related to demand, such as errors in the O-D matrix.

Using data from the O-D matrix and considering roadway network characteristics, one can determine the route choices the traveler will make. So, a calibration parameter at the network level can be the outcome of traffic assignment, which is the route choices travelers will make. The simplest route choice selection is based on travel time (or travel cost). The volume of traffic on a network link would depend on the route choices assigned to the travelers. Thus, it directly affects how the network is filled with traffic. Barcelo (Chapter 9 of Fundamentals of Traffic Simulation, Ed. by Barcelo, 2010) and Casas (2006) proposed some practical approach to deal with route choice issues at the network level and how traffic is assigned to the link.

6.3 Model Verification

Verification is to be sure that the model will behave as intended. The program should be debugged first to eliminate any coding errors and programming problems. Then, the logic of different components of the model, such as car following, lane changing, merging, or diverging, should be carefully reviewed. Also, the acceleration and deceleration patterns, velocity change patterns, trajectory plots, and headways obtained from the simulation model should be examined.

To illustrate the verification concept, we used the CARSIM (Benekohal & Treiterer, 1989; Benekohal, 1991) simulation model. Different disturbances in traffic flow were induced in a platoon of 15 vehicles, and their effect at microscopic and macroscopic levels was examined. A disturbance is induced when the leader of a platoon is required to decelerate, stop, and accelerate to a specified speed.

For microscopic level verification, the effects of a regular disturbance, an emergency stop, and a stop-and-go operation on individual vehicle's trajectory, speed, and acceleration or deceleration were analyzed at high and low volume levels. For the macroscopic level verification, the effects of the regular disturbance and an emergency deceleration on average speed, density, volume, and average headway were examined at high and low volume levels. The acceleration and deceleration patterns for a 15-car platoon in a stop-and-go condition are shown in Figure 6.2. The leader of the platoon decelerates at 16 ft/sec/sec, stops for 9 sec, and accelerates to a desired speed. The patterns are shown for the first car, the last car, and every third car in the platoon. It shows that the disturbance is dampening, and traffic flow is stable, as expected in this case.

6.4 Model Validation

6.4.1 *Validation at What Level*

To validate the simulation results, one needs to compare the simulation output to a more reliable source of information, such as reliable field data. One issue that needs to be decided is at what level the validation would be performed. It can be performed at the following levels:

- network level;
- arterial level;
- link level (a street segment);
- node (Intersection) level.

These levels are schematically shown in Figure 6.3.

At the network level, one may look at the total trips completed in the network, trips completed at each exit point, all trips ending on the exits to the east (west, north, south), on selected key exit points, on certain key O-D pairs, or along a certain path. More on this is discussed in the macroscopic validation of DTA models.

At the arterial level, one may look at certain traffic characteristics on that arterial, such as average travel time or speed, exit volume, density, signal progression, and arterial delay. When there are more arterials under study, the analyzed arterial should be selected carefully so that it represents the other arterials.

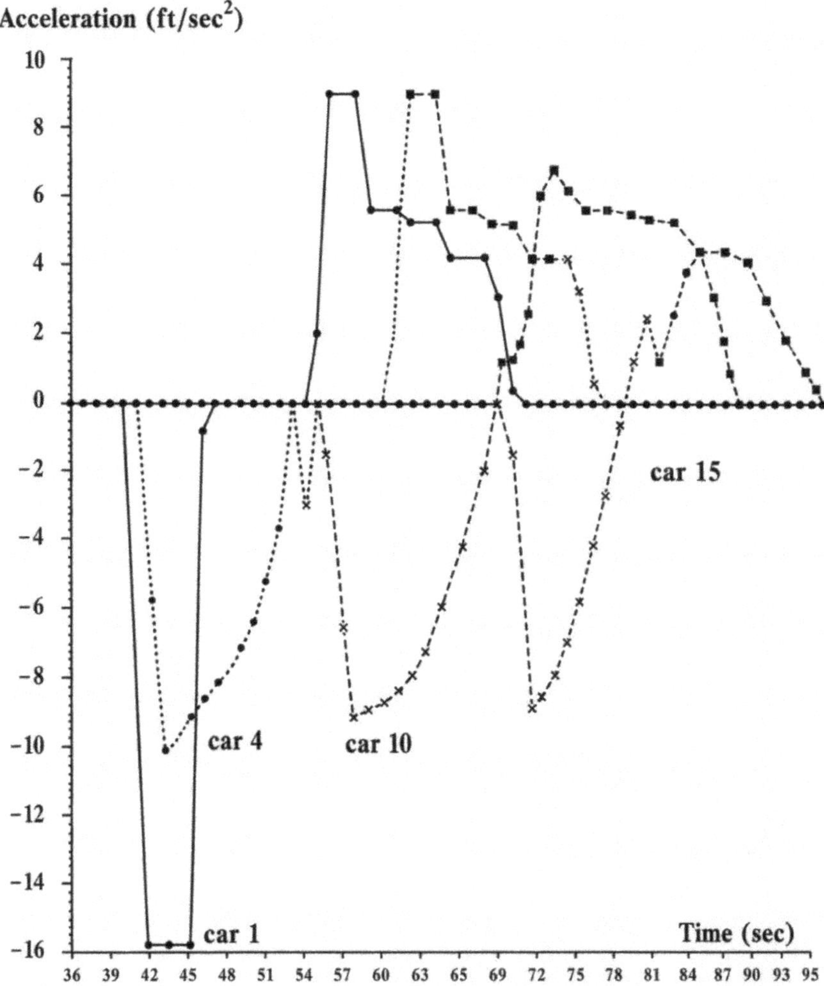

Figure 6.2. Acceleration-deceleration patterns for a platoon of vehicles in a stop-and-go condition. The leader of the platoon decelerates at 16 ft/sec/sec, stops for 9 sec, and accelerates to a desired speed. (From Benekohal, R. F. Procedure for Validation of Microscopic Traffic Flow Simulation Models. Transportation Research Record: Journal of the Transportation Research Board, No. 1320, 1991, Figure 1, p. 193. Copyright, National Academy of Sciences. Reproduced with permission of the Transportation Research Board).

At the link level, one may look at exit volume, speed, density, travel time, etc. At the node level, one may look at delay per vehicle, number of users entering/leaving the node (intersection), no of stops, comfort level, level of service, etc.

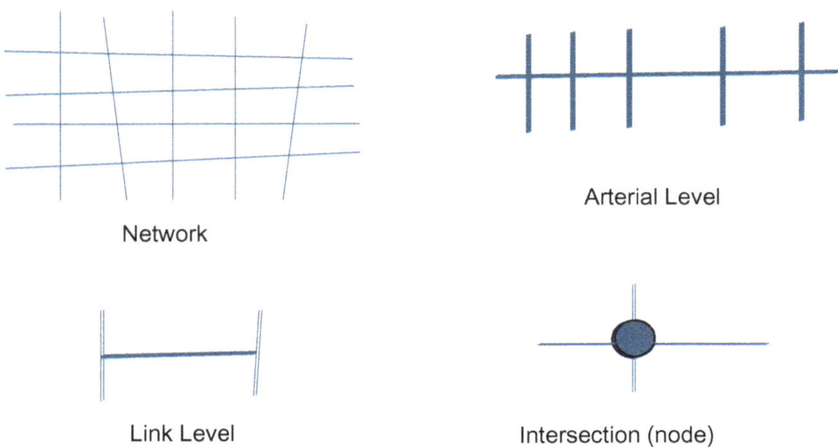

Figure 6.3. Data collection levels.

6.4.2 *Techniques Used in Model Validation*

Various statistical techniques and traffic flow variables have been used for the validation of simulation models. Using chi-squared, factor analysis, spectral analysis, and regression analysis was discussed in the literature (Kleijnen, 1975b). Also, using three alternative forms of analysis of variance was presented (Naylor *et al.*, 1967). Using regression analysis to obtain a metamodel (a model explaining the simulation model) and how the effect of qualitative or quantitative factors can be investigated by using the weighted least squares technique was discussed by Kleijnen (1982). The procedures for comparison of the results from several independent replications of a simulation model with that of one or more observed data were discussed in Mihram (1972), and Kleijnen (1976, 1977).

The variable to be used in validation mainly depends on the data available or can be easily collected. Normally, volume and speed data from loop detectors in the field are compared to volume and speed data collected from virtual loop detectors placed at the same locations in the simulation model. Travel time, which is a derivative of speed, may also be used in validation. For example, travel time and velocity of vehicles were used as measures of effectiveness, and the simulation results were compared with observed values using the Wilcoxon signed-rank test (Gafarian & Walsh, 1970).

Two levels are proposed for validation of a microsimulation model:

(1) microscopic level;
(2) macroscopic level.

At the microscopic level, attributes of individual vehicles such as location, time, headway, and speed computed from the simulation model are compared with those from field data. At the macroscopic level, aggregate parameters such as average speed, density, and volume computed from the simulation model are compared with the results from field data. Schematically, the steps in the validation of the car following model (in this case, CARSIM) are shown in Figure 6.4.

6.4.3 *Validation at Microscopic Level*

Microscopic level validation is, perhaps, the most challenging step in the validation of a microsimulation model. For microscopic validation, the variation of some or all of the following parameters should be examined:

- speed profile of an individual vehicle;
- location of vehicle on the road (trajectory);
- time headways between vehicles;
- spacing between successive vehicles.

The illustration of how the first two items can be used is presented in the following sections.

6.4.3.1 *Speed Profile Comparison*

The speed profile shows the speeds of an individual vehicle over time. At short time intervals (e.g. 1 sec), speed for each vehicle is generated and plotted vs time. If replication runs are made, the average speed of a vehicle at a given time is computed as the mean of the speeds from each replication. Then, the average speed of an individual vehicle at a given time was compared to the observed speed from field data. Comparisons of the speeds from simulation and field data are shown in Figure 6.5, where the speeds for the first vehicle, the last vehicle, and every third vehicle are plotted.

As can be observed, the simulation model generated speed profiles like those from the field data. It is important to note that all vehicles came to a complete stop and then accelerated to reach their desired speed.

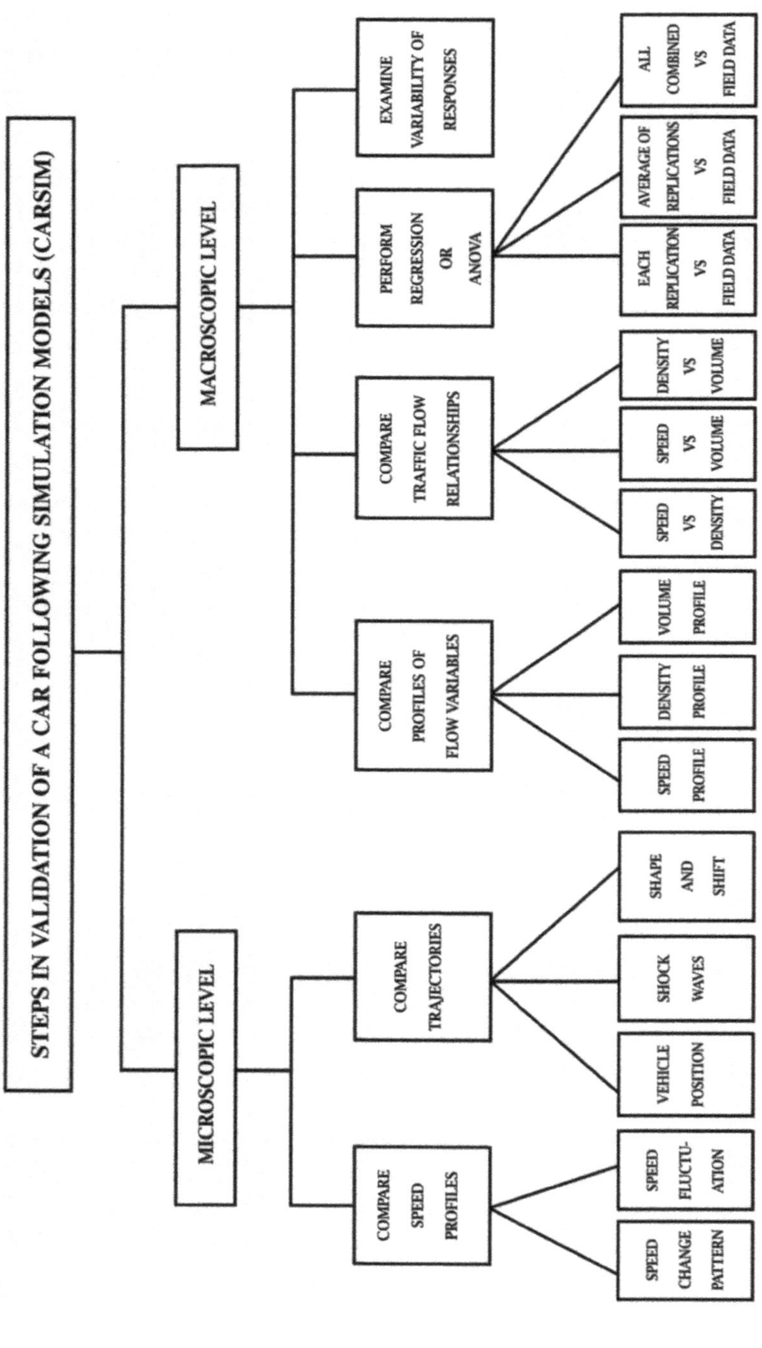

Figure 6.4. Steps in validation of the car-following simulation model. (From Benekohal, R. F. Procedure for Validation of Microscopic Traffic Flow Simulation Models. Transportation Research Record: Journal of the Transportation Research Board, No. 1320, 1991, Figure 2, p. 194. Copyright, National Academy of Sciences. Reproduced with permission of the Transportation Research Board).

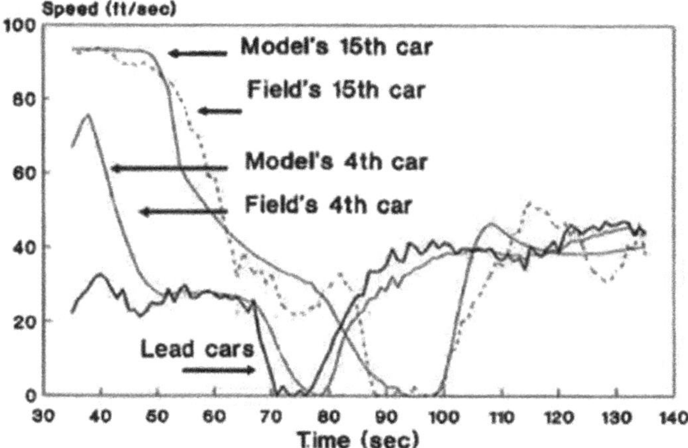

Figure 6.5. Comparison of speed change patterns from simulation vs field data for the first, fourth, and last cars in Platoon 123. (From Benekohal, R. F. Procedure for Validation of Microscopic Traffic Flow Simulation Models. Transportation Research Record: Journal of the Transportation Research Board, No. 1320, 1991, Figure 3, p. 194. Copyright, National Academy of Sciences. Reproduced with permission of the Transportation Research Board).

The similarity of the speed change patterns indicates that the simulation model replicates the real-world traffic disturbance reasonably well. In the simulated platoon, the vehicles showed less speed fluctuation than the vehicles in the field data because the simulation plots are the average of several replication runs.

From the speed profile comparison, one may conclude how well the model duplicates the real-world speed change patterns. The criteria for the operational validation may be the comparison of the shape of the speed profile and the shift between the two profiles. A driver with a longer reaction time would cause a larger shift than a driver with a short reaction time, but the shape of the profile for both drivers may look similar.

6.4.3.2 *Trajectory Comparison*

A vehicle trajectory is obtained by plotting the position of the vehicle vs the time (often in 1 sec time intervals). The average of the replications is used as the location of the vehicle at a given time. The trajectory plots for every third vehicle including the last vehicle are shown in Figure 6.6.

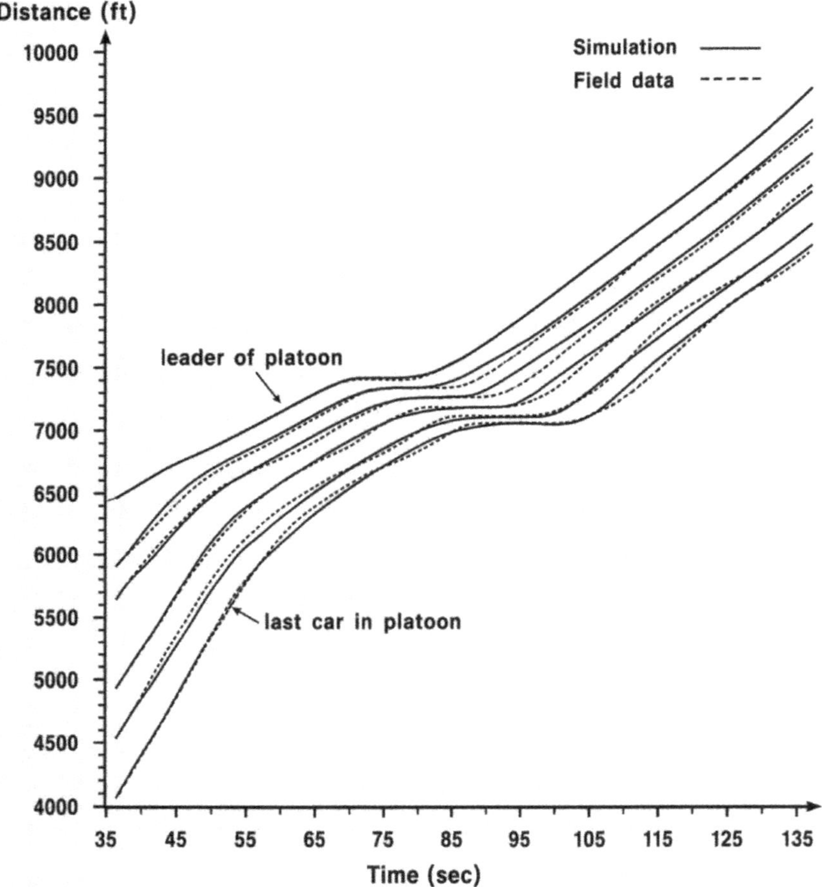

Figure 6.6. Comparison of vehicles trajectories from the simulation model versus field data for Platoon 123. The trajectories are shown for every third car in a platoon of 15 cars. (From Benekohal, R. F. Procedure for Validation of Microscopic Traffic Flow Simulation Models. Transportation Research Record: Journal of the Transportation Research Board, No. 1320, 1991, Figure 4, p. 195. Copyright, National Academy of Sciences. Reproduced with permission of the Transportation Research Board).

When there is a severe disturbance in the traffic flow, it is quite challenging for a simulation model to generate trajectories that are close enough to the actual trajectories. However, in normal traffic conditions, it is not very difficult to obtain trajectory plots that are close to the trajectory plots from field data. Thus, models validated only in free-flow traffic conditions may not accurately simulate high-density traffic conditions. Also, validated models using data only from either the acceleration or the deceleration phase of a traffic disturbance may not accurately simulate stop-and-go conditions.

The criteria to evaluate the similarity between the model and field data may include vehicle location, difference in location between the model and real platoon, shape of the plot for individual vehicles, general shift up or down, shift either before or after the disturbance, and location at which a vehicle slows down, stops, starts, and recovers. It is important to use short time intervals (a few seconds) in generating trajectory plots if the model will be used for detailed studies. Otherwise, the model may not accurately show the behavior of traffic within that interval. For instance, Figure 6.6 shows that the vehicles in the platoon stopped and moved in less than 20 sec. If the data had been collected every 30 sec, the stop-and-go behavior of the platoon would have been missing.

The microscopic validation may be conducted subjectively or objectively, as suggested by Sargent (1982). The graphical comparison of (subjective validation) the speed profiles and trajectory plots was found to be sufficient at this level. The objective validation (statistical technique) was not used because of a strong correlation between successive points on the speed profile or the trajectory plots for a vehicle. Once satisfactory results were obtained from the microscopic level validation, the macroscopic validation was started.

6.4.4 *Validation at Macroscopic Level*

At the macroscopic level, aggregate parameters such as average speed, density, and volume computed from the simulation model are compared with field data. For macroscopic validation, the traffic stream parameters are used instead of individual vehicle characteristics. The macroscopic level validation may not reveal as much information about model capability as the microscopic level because the variables are the average of all vehicles. For instance, the average speeds of simulated and field data may be similar enough, but the speed of individual vehicles may still be very different. Likewise, the density might be very close to that of field data, but the spacing between successive vehicles may be considerably different.

For the macroscopic level validation, the following comparisons are suggested:

1. comparison of the profile of traffic flow variables;
2. comparison of fundamental relations of traffic flow;
3. comparison of simulation results vs field data.

In addition to the comparisons, the range of variations of the response variables should also be examined. Application of these concepts in the validation of CARSIM is discussed in the following sections.

6.4.4.1 *Comparison of Profile of Traffic Flow Variables*

Traffic flow variables used for the comparison were speed, density, and volume. These variables were computed at 1 sec time intervals, and their variations over time (profile) were compared to the field data, as shown in Figure 6.7. The platoon traveling at a speed of more than 80 ft/sec reached a speed of near zero in less than 1 min. The simulation results are very close to the actual traffic speeds.

The plots of density vs time for Platoon 123 and the simulation counterparts are shown in Figure 6.8. The graphs show the same patterns and fluctuations for both simulation and actual platoons. The time a simulated platoon reaches the jam density is very close to that of the actual platoon. The density of a platoon is computed from the distance between the first and the last car in the platoon. The distance is very dependent on the spacing between these two cars. Therefore, one should be very careful in using the density of a platoon for comparison of actual and simulated results.

Comparison of the volume profile for the simulated and actual platoon is shown in Figure 6.9. The volume is computed as the product of speed and density. The difference between the simulated and actual volume may be large due to the multiplication. Another reason for the large difference might be that the volume may not be equal to the product of speed and density when traffic flow breaks down (critical density). Thus, the volume comparison is not recommended when traffic flow density reaches its critical range.

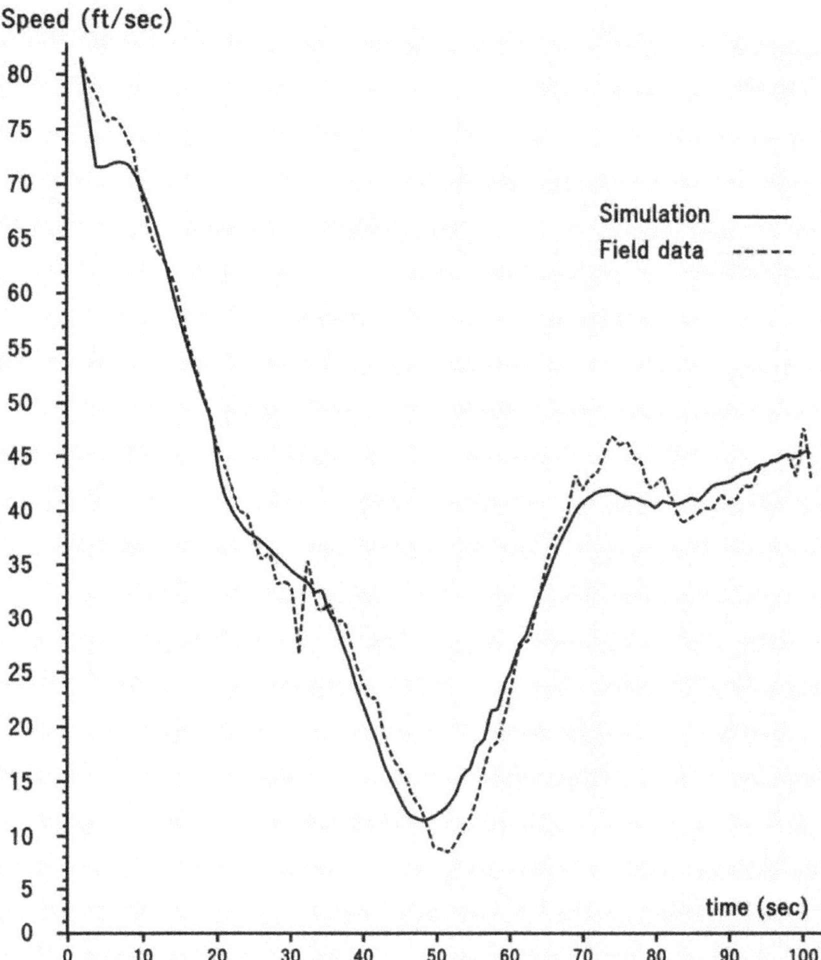

Figure 6.7. Comparison of the average speed from the simulation model versus field data for Platoon 123. (From Benekohal, R. F. Procedure for Validation of Microscopic Traffic Flow Simulation Models. Transportation Research Record: Journal of the Transportation Research Board, No. 1320, 1991, Figure 5, p. 196. Copyright, National Academy of Sciences. Reproduced with permission of the Transportation Research Board).

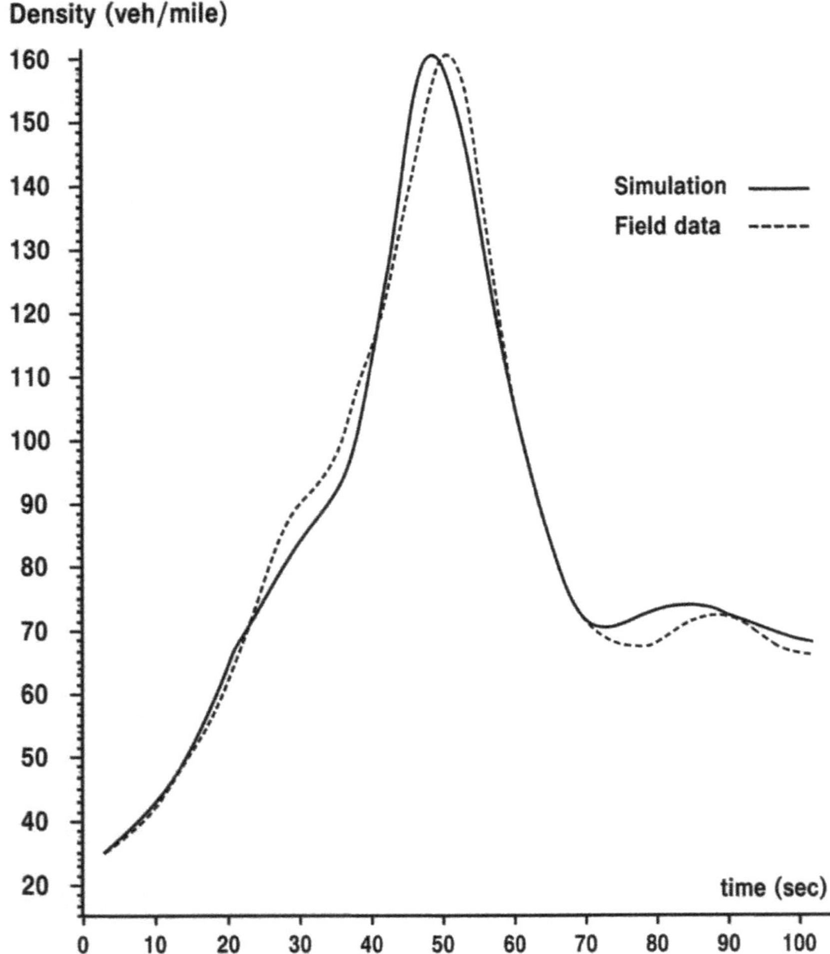

Figure 6.8. Comparison of density from the simulation model versus field data for Platoon 123. (From Benekohal, R. F. Procedure for Validation of Microscopic Traffic Flow Simulation Models. Transportation Research Record: Journal of the Transportation Research Board, No. 1320, 1991, Figure 6, p. 197. Copyright, National Academy of Sciences. Reproduced with permission of the Transportation Research Board).

6.4.4.2 *Fundamental Relations of Traffic Flow*

The fundamental relationships among traffic flow parameters (speed, density, and volume) from simulation are compared to field data. Speed–density, speed–volume, and density–volume relationships from the field

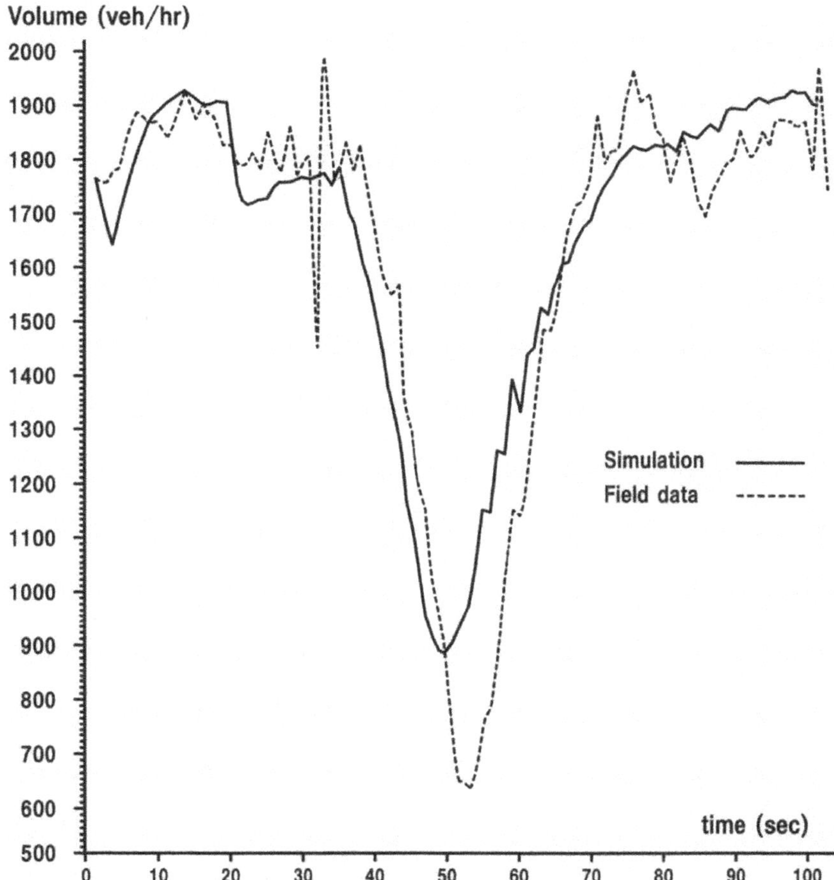

Figure 6.9. Comparison of volume from the simulation model versus field data for Platoon 123. (From Benekohal, R. F. Procedure for Validation of Microscopic Traffic Flow Simulation Models. Transportation Research Record: Journal of the Transportation Research Board, No. 1320, 1991, Figure 7, p. 198. Copyright, National Academy of Sciences. Reproduced with permission of the Transportation Research Board).

data are compared to those from the simulation model. The speed–density relationships for simulated and field data are presented in Figure 6.10. The actual data show a nonlinear relationship between speed and density, and a loop representing the hysteresis phenomenon when the traffic flow breakdown occurs (Treiterer, 1975). The simulation results show similar relationships and the same phenomenon.

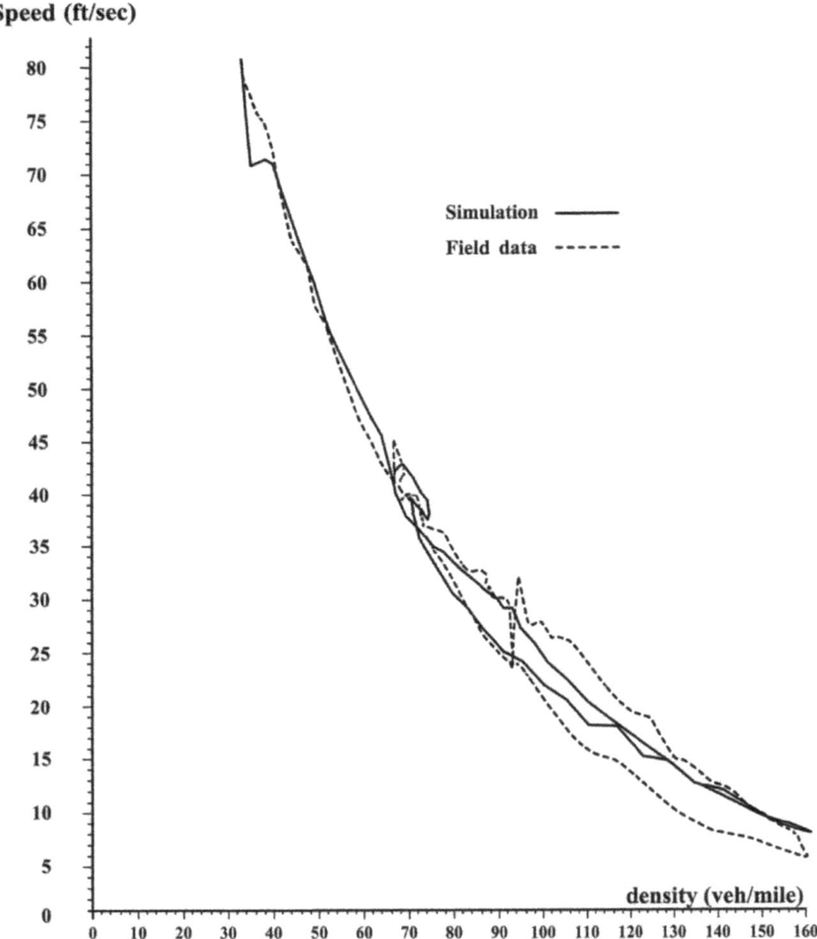

Figure 6.10. Comparison of speed-density relationships from the simulation model versus field data for Platoon 123. (From Benekohal, R. F. Procedure for Validation of Microscopic Traffic Flow Simulation Models. Transportation Research Record: Journal of the Transportation Research Board, No. 1320, 1991, Figure 8, p. 199. Copyright, National Academy of Sciences. Reproduced with permission of the Transportation Research Board).

Similar comparisons should be made with speed–volume and volume–density relationships. In addition to the graphical presentation of the results, the statistical analysis of the simulation results vs the actual data is also carried out. Statistical analysis is discussed in the following section.

6.4.4.3 *Comparison of Traffic Flow Variables from Simulation vs Field Data*

Traffic flow variables (speed, density, and flow) obtained from the simulation model should objectively (e.g. statistical analysis) be compared to field data. Statistical techniques such as regression analysis, analysis of variance, or time series analysis may be used for such comparisons. The appropriate method should be selected based on the type of data available. Application of regression analysis and analysis of variance (ANOVA) for a simulation model performance evaluation is discussed here.

Regression analysis of speed, density, and volume computed from the simulation model vs those from the field data is illustrated here. For each time interval, the average speed, density, and volume were computed from simulation and field data. The general form of the regression lines is

$$P_{\text{model}} = b_0 + b_1 * (P_{\text{field}})$$

where P_{model} is a parameter (such as speed, density, or volume) from the simulation model, P_{field} is a parameter (such as speed, density, or volume) from the field data, b_0 is the y-intercept of the regression line, and b_1 is the slope of the regression line.

Independent replications must be made for any simulation model with a stochastic parameter. One must be very careful in interpreting the differences among the replication results. The following comparisons may be considered among independent replications:

1. comparison of an individual replication vs field data;
2. comparison of the average of replications vs filed data;
3. comparison of all replications combined vs field data.

6.4.4.3.1 Comparison of an Individual Replication vs Field Data

For comparison of individual replication vs field data, speed, density, and volume computed from each simulation run are regressed over the values from field data. The coefficients of the regression lines, variance of the coefficients, and R^2 values are summarized in Table 6.1. Note that $s(b0)$ and $s(b1)$ are the variances of b_0 and b_1. The R^2 values and the coefficients of the regression lines indicate that there is a strong agreement between the simulation results and the field data. It should be noted that the slope and y-intercept of the regression lines, as well as the $R2$ values for a given

Table 6.1. Regression of speed, density, and volume from individual simulation runs vs field data.

Replications	*b*0	*b*1	*R***2	*S*(*b*0)	*S*(*b*1)
SPEED					
1	1.09696	0.96812	0.98129	0.54144	0.01344
2	1.96750	0.94091	0.98012	0.54266	0.01348
3	1.91585	0.93815	0.98287	0.50152	0.01245
4	2.44674	0.91579	0.98478	0.46103	0.01144
5	1.10296	0.97226	0.97963	0.56778	0.01409
DENSITY					
1	1.90606	0.98211	0.98744	0.95538	0.01113
2	2.05298	0.95981	0.98826	0.90241	0.01051
3	1.40735	0.97494	0.99118	0.79343	0.00924
4	2.32060	0.94958	0.99130	0.76726	0.00894
5	–0.49143	1.05514	0.97827	1.35641	0.01580
VOLUME					
1	353.88146	0.79593	0.81033	64.38405	0.03870
2	277.90502	0.82216	0.82465	63.38825	0.03810
3	415.23342	0.74210	0.82827	56.49940	0.03396
4	383.34560	0.74492	0.88891	44.03049	0.02647
5	316.72237	0.85828	0.79982	71.79334	0.04315

Source: From Benekohal, R. F. Procedure for Validation of Microscopic Traffic Flow Simulation Models. Transportation Research Record: Journal of the Transportation Research Board, No. 1320, 1991, Table 1, p. 200. Copyright, National Academy of Sciences. Reproduced with permission of the Transportation Research Board).

parameter, do not show a large variation among the replications. When the variation is large, more replications should be made.

The advantage of using regression of individual runs over using the average of the five replications is that one would get additional information about the variation of the response variables among the replications. When the average values are used, this information is no longer available; however, the results are easy to interpret.

6.4.4.3.2 Comparison of Average of Replications vs Field Data

For comparing the average of replications vs field data, the speed, density, or volume at a given time is computed as the average value from five

Table 6.2. Parameters of regression lines for the average of five replications and all five replications combined vs the field data.

Type of Analysis	Variable	*b*0	*b*1	*R***2	*s*(*b*0)	*s*(*b*1)
Average of five runs	Speed	1.70554	0.94705	0.98383	0.49177	0.01220
	Density	1.43827	0.98432	0.98922	0.88620	0.01033
	Volume	349.433	0.79265	0.84406	56.9654	0.03424
All five runs combined	Speed	1.70655	0.94705	0.98100	0.23682	0.00588
	Density	1.44075	0.98432	0.98123	0.52091	0.00607
	Volume	349.472	0.79267	0.80913	28.5566	0.01717

Source: From Benekohal, R. F. Procedure for Validation of Microscopic Traffic Flow Simulation Models. Transportation Research Record: Journal of the Transportation Research Board, No. 1320, 1991, Table 2, p. 200. Copyright, National Academy of Sciences. Reproduced with permission of the Transportation Research Board).

replications. Then, the average of five replications is compared to field data. Table 6.2 shows some of the parameters obtained from the regression analysis. The slopes of the regression lines are very close to one, and the *y*-intercepts are very close to zero. The *R*-Squared values for speed and density are 0.98 or higher, and for volume, it is 0.80 or higher. The results indicate that there is a strong agreement between the speeds or densities computed from the simulation and the field data. As was discussed before, the agreement for the traffic volumes is as strong as speed or density.

When the average values of several replications were compared to field data, the parameters showed less variation than the same parameter in the individual runs. Although using the average values of the replications makes the comparisons less complicated, finding the average of several replications may conceal very useful information about the sensitivity of a parameter to an input variable. For instance, the speed regression lines in Table 6.1 show the range of variation of slope, but the speed regression line in Table 6.2 does not. On the other hand, since there are five values from the simulation model for each value from the field data, one might treat them as repeated observations at a given point. The consequence of this assumption is discussed in the following section.

6.4.4.3.3 Comparison of All Replications Combined vs Field Data

For comparison of all replications combined vs field data, the values obtained from the individual simulation runs are combined to create one dataset. Regression analysis of the combined dataset vs field data yielded, as it was expected, slopes and y-intercepts equal to that of the average of five replications, see Table 6.2. However, the variance of slope and y-intercept decreased almost to one-half of that of the average of five replications. There was also a slight decrease in the $R2$ values.

It should be noted that the assumption about the repeated data was proved to be incorrect. In regression analysis, when there are repeated observations at a given point, the "lack of fit" of the regression line should be examined (Draper & Smith, 1981). The examination of ANOVA tables falsely indicated the lack of fit for the linear model.

The possibility of fitting a higher-order model was explored, and the residuals were plotted vs the time and predicted values. The plots did not show any definite trend. The residuals were clustered around the line $y = 0$ and approximately made a horizontal band. The lack of the trend, presence of the band, and $R2$ value very close to one indicated the adequacy of the model. Thus, there was no lack of fit.

The lack of a fit test is not appropriate for this situation because there is a strong dependency between successive data points. For instance, the density currently interval depends on the density of the previous time interval. The false detection of lack of fit was not due to the inadequacy of the linear regression model but due to using the lack of fit test when the data points are strongly correlated.

6.4.4.3.4 Variation of the Response Variables

The result obtained from a single simulation run may not be very reliable when stochastic variables in the model affect the model's outcome. One method to increase reliability and confidence in the simulation responses is to make several independent replications. Then, the change in the response variable among the replications should be examined. The speed variation in CARSIM is shown in Figure 6.10, and it shows very small variation from one simulation run to another. When a wider variation is observed, more runs should be made.

6.4.5 *Calibration and Validation of Commercial Software*

6.4.5.1 *VISSIM Calibration and Validation*

VISSIM is a microscopic traffic simulation model, and more details about its model development and validation are given by Fellendorf and Vortisch

(2010). For calibration of VISSIM, Menneni *et al.* (2008) used speed–flow relationship. They used NGSIM trajectory data (Kovvali *et al.*, 2007) from Interstate 101 in California and genetic algorithms to change the driver behavior parameters until the field and simulation data were close enough. In a subsequent study (Menneni *et al.*, 2009), they used trajectory data to further calibrate Wiedemann's car following model parameter to US driving conditions.

6.4.5.2 *S-Paramics Calibration and Validation*

S-Pramics (sometimes referred to as Paramics) is a microsimulation model and assigns certain characteristics to drivers and vehicles to represent observed behaviors (Skyes, 2010). For calibration of vehicle behavior, it is suggested (Skyes, 2010) to look at (a) network-wide behavior and (b) individual junction/link behaviors. It says that network-wide behavior normally should not be calibrated unless you believe that divers behave differently in the entire modelled area. The key overall driver behavior parameters are vehicle aggression and awareness distribution, and network headway factors. For calibration at the junction level, it compares observed and modeled queue lengths at a junction. The key calibration parameter is visibility, which indicates if a turning vehicle will come to a complete stop. Link attributes (width, number of lanes, curvature, speed limit, and lane restrictions) may be considered for calibration, if needed. There are also gap acceptance parameters for lane merging, lane crossing, and path crossing, and headways may be changed only if there are special circumstances or enough support from local observed data. For validation, it suggested looking at link and turn flow counts, travel time, and queue length.

6.4.5.3 *Aimsun Calibration and Validation*

AIMSUN (Advanced Interactive Microscopic Simulator for Urban and non-urban Networks) originally started as a microscopic simulation model and then expanded to macroscopic and mesoscopic models and is referred to as Aimsun (Casas *et al.*, 2010). Which parameters should be calibrated depend on the type of model (microscopic, mesoscopic, and macroscopic), objective of the study, and type of network (Casas *et al.*, 2010). The most significant measures they proposed to be calibrated for highway models are the relationship between speed/flow/density, lane utilization, and congestion propagation; for urban models, queue length, queue discharge rate, and level of service; and for large and/or complex networks travel time and traffic flow parameters. For validation, comparison may be made for aggregated measurements (e.g. hourly flow at counting stations), time series analysis of an output variable, or band analysis of the variable that shows the range of changes.

6.4.5.4 *MITSIMLab Calibration and Validation*

MITSIMLab (microscopic traffic simulation laboratory) is a microscopic traffic simulation model developed by MIT and was used to evaluate alternative traffic management strategies for the Central Artery/Tunnel in Boston (Ben-Akiva *et al*., 2010). It has been used as a laboratory/tool to evaluate ITS and other traffic management strategies worldwide. The structure of models in MITSIMLab formed the basis for TransModeler software (Ben Akiva *et al*., 2010). Driver acceleration/deceleration value is computed based on an adapted GM model (Gazis *et al*., 1961), which is based on the stimulus–response concept for car following. Calibration and validation of MITSIMLab are performed at disaggregate and aggregate data levels. In the disaggregate calibration, the behavior model components (e.g. acceleration, lane changing, and route choice) at the individual level are estimated using detailed data. This calibration activity yields the estimated models that will be used in the aggregate calibration. In the aggregate calibration phase, other model components and the estimated models are jointly calibrated using aggregate data, such as volume counts or average speed. Part of the aggregate dataset is used to adjust the key parameters of the behavior model. The aggregate calibration is formulated as an optimization problem where the goal is to minimize the deviation of the simulated and observed traffic measurement and the deviation of the calibrated values from their prior estimates, if available (Ben-Akiva *et al*., 2010). For validation, performance measures (e.g. Average speed, flow, lane distribution, intensity, and location of lane changes) computed by the simulation model are compared to the observed field data. For validation, MITSIMLab in Stockholm, Sweden Study (Toledo *et al*., 2003), they used traffic flow, travel times, and queue length during 2 hr of AM peak data. Traffic flows were also used to estimate OD matrices.

6.4.5.5 *SUMO Calibration and Validation*

SUMO (Simulation of Urban Mobility) is an "open-source" microscopic simulation program. Originally developed in Germany, SUMO does not specify what parameters to be calibrated but emphasizes the role of verification in obtaining meaningful outcomes. It used the car following model developed by Krauss (1998), where the follower is moved forward with an acceleration rate that avoids collision with its leader. SUMO assumes that drivers are not perfect in realizing that they have reached their desired speed, so there is a driver imperfection parameter that is implemented by assigning a stochastic value between 0 and 1. The acceleration parameter and the driver imperfection distribution may be used that can be calibrated.

6.4.6 *Simulation Manuals Developed by DOT*

The FHWA (Federal Highway Administration) office of the US DOT has publications about the use of traffic simulation models FHWA (2004), and FHWA (2007). Some state DOTs, such as Wisconsin DOT, have developed their own manuals. The state DOT manuals often are for specific software that the DOT uses. An example of such a manual is that of Wisconsin DOT (WisDOT). It requires both quantitative and qualitative validations of the model outputs. It describes the step-by-step tasks that need to be completed and gives the validation metrics and acceptance thresholds required. Examples of these are given in Tables 6.3–6.5. It requires a tiered validation approach where "the Tier 1 test would be a global validation test for a metric, and the Tier 2 test would be a local test for that same metric. If a model passes the Tier 1 (global) test, the modeling team would not need to perform the Tier 2 (local) test, and a detailed summary of the Tier 2 test would not be necessary". The validation MOEs and acceptance thresholds for the Tier 1 (global) test are given in Table 6.3.

The Tier 1 (global) validation tests, shown in Table 6.4, are applicable for link/segment volumes, travel times, and travel speeds. If the model does not pass the Tier 1 tests, the Tier 2 (local) test shall be performed. The Tier 2 MOEs and validation acceptance threshold are given in Table 6.5.

Table 6.3. Parameter specified by Wisconsin DOT for Validation Tests.

MOE	Criteria	Validation Acceptance Threshold	
Volume[(a)]	All Links > 100 vph (Mainline and Critical[(b)] Arterials)	Tier 1:	RMSPE <5.0%
		Tier 2:	RNSE <3.0% for >85% of links
	All Turns	Tier 1:	Not Applicable
		Tier 2:	RNSE <3.0% for >75% of turns
Speeds	All Segments or Spot-Speed Locations	Tier 1:	RMSPE <10.0%
		Tier 2:	Within ± (Mainline Posted Speed X 20%) for >85% of locations
Travel Times	All Routes > 1.5 Miles	Tier 1:	RMSPE <10.0%
		Tier 2:	Within ± 15% for >85'% of routes
Queues	All Critical[(b)] Queue Locations	Tier 1:	Not Applicable
		Tier 2:	± 150 feet for queues 300 to 750 long, Within ±20% for queues >750 feet long
Lane Use	All Critical[(b)] Lane Utilization Locations	Tier 1:	Not Applicable
		Tier 2:	RNSE <3.0% for >85% of locations consistent with field conditions

Notes:

[(a)] All traffic models **shall** undergo volume validation (Tier 1) tests

[(b)] Critical locations are those locations likely to have an impact on operations to the project study area (e.g., locations with higher traffic volumes, existing or projected level of service is at or approaching unstable flow, queues block or impede travel, weaving areas, merge/diverge locations, etc.)

vph = vehicles per hour

RMSPE = Root Mean Squared Percent Error, See TEOpS 16-20-8.4 for equation

RNSE = Root Normalized Squared Error, See TEOpS 16-20-8-5.1 for equation

Reproduced with permission of Wisconsin Department of Transportation.

The Wisconsin DOT manual requires that qualitative validation checks be performed in addition to quantitative tests. Table 6.6 (Table 8.4 Wis DOT) of the Wisconsin DOT provides a summary of the qualitative validation checks. It says, "The analyst shall perform the qualitative validation tests for all models, even those that pass the Tier 1 (global) mathematical validation thresholds". The outcome of this qualitative validation should be documented and the decision made justified.

Table 6.4. Global parameter specified by Wisconsin DOT for Tier 1 (Global) Validation Tests.

MOE	Criteria	Validation Acceptance Threshold	
Volume[(a)]	All Links > 100 vph (Mainline and Critical[(b)] Arterials)	Tier 1:	RMSPE <5.0%
Speeds	All Segments or Spot-Speed Locations	Tier 1:	RMSPE <10.0%
Travel Times	All Routes > 1.5 Miles	Tier 1:	RMSPE <10.0%

Notes:

[(a)] All traffic models **shall** undergo volume validation (Tier 1) tests

[(b)] Critical locations are those locations likely to have an impact on operations to the project study area (e.g., locations with higher traffic volumes, existing or projected level of service is at or approaching unstable flow, queues block or impede travel, weaving areas, merge/diverge locations, etc.)

vph = vehicles per hour

RMSPE = Root Mean Squared Percent Error, See TEOpS 16-20-8.4 for equation

Reproduced with permission of Wisconsin Department of Transportation.

Table 6.5. Local parameter specified by Wisconsin DOT for Tier 2 (Local) Validation Tests.

MOE	Criteria	Validation Acceptance Threshold	
Volume[(a)]	All Links > 100 vph (Mainline and Critical[(b)] Arterials)	Tier 2:	RNSE <3.0% for >85% of links
	All Turns	Tier 2:	RNSE <3.0% for >75% of turns
Speeds	All Segments or Spot-Speed Locations	Tier 2:	Within ± (Mainline Posted Speed X 20%) for >85% of locations
Travel Times	All Routes > 1.5 Miles	Tier 2:	Within ±15% for >85% of routes
Queues	All Critical[(b)] Queue Locations	Tier 2:	± 150 feet for queues 300 to 750 long, Within ±20% for queues >750 feet long
Lane Use	All Critical[(b)] Lane Utilization Locations	Tier 2:	RNSE <3.0% for >85% of locations Consistent with field conditions

Notes:

[(a)] All traffic models that do not pass the Tier 1 validation test **shall** undergo the Link/Segment Volume Tier 2 validation tests. All traffic models that include intersections **shall** undergo the Turning Volume Tier 2 validation tests.

[(b)] Critical locations are those locations likely to have an impact on operations to the project study area (e.g. locations with higher traffic volumes, existing or projected level of service is at or approaching unstable flow, queues block or impede travel, weaving areas, merge/diverge locations, etc.)

vph = vehicles per hour

RNSE = Root Normalized Squared Error, See TEOpS 16-20-8-5.1 for equation

Reproduced with permission of Wisconsin Department of Transportation.

Table 6.6. Qualitative Validation tests suggested by Wisconsin DOT for Quantitative Validation Tests.

MOE	Criteria	Validation Acceptance Threshold
Queues	All Critical Queue Locations	Visually realistic for intersection queues. Quantitative checks required if queues are a primary validation MOE.
Bottlenecks	Replication of Real-World Bottlenecks	Visually realistic for intersection queues and freeway bottlenecks.
Routing	All Routes	Represents field conditions and driver behavior. Acceptance of quantitative results require WisDOT approval.
Lane Use	All Critical Lane Utilization Locations	Visually realistic. Quantitative checks encouraged for areas where lane usage has a considerable influence on operations.
Freeway Merging	All Merge Locations	Visually realistic.
Vehicle Types and Truck Percentages	All Locations	Represents field conditions.

Exercises

1. Why do we calibrate simulation models?
2. What are we trying to achieve by the verification of simulation models?
3. What is the main issue with visual validation?
4. Why should one not use the same data for calibration and validation?
5. Why is it a good idea to validate simulation models at microscopic and macroscopic levels?
6. List the important variables that can be calibrated at the microscopic level (driver–vehicle unit level).
7. If a model is not valid at the microscopic level but happens to be valid at the macroscopic level, would you consider it a valid model?
8. Define calibration, verification, and validation in the context of traffic simulation. Explain how they differ in purpose and timing.
9. Why should validation not be treated as an "either/or" proposition? What does it mean to tolerate discrepancy?

10. What is accreditation? How is it different from calibration and validation?
11. Explain the difference among conceptual validation, computerized validation, and operational validity. Provide one example task for each.
12. Define face validation and tracing. Why are they useful for microscopic models?
13. Explain overfitting in calibration. Why can overfitting increase validation error rates?
14. A model matches mean travel time well but produces unrealistic lane changing and shockwave behavior. Is this primarily a calibration issue, verification issue, validation issue, or multiple? Explain.
15. You validated a model only under free-flow conditions. Explain two risks when applying it to congested conditions.
16. Create a verification checklist with at least 12 items covering geometry/network coding, signal control plans, demand inputs, car following checks, lane changing checks, and output/runtime checks.
17. Explain why macroscopic validation may hide errors. Provide a concrete example comparing averages vs individual vehicle behavior.
18. This chapter warns that volume comparison may be unreliable near critical density. Explain why and propose what you would compare instead.
19. Compare three validation approaches with replications: (1) individual replication vs field, (2) average of replications vs field, and (3) all replications combined vs field. For each, state one advantage and one risk/limitation.
20. Explain why "lack of fit" tests can be misleading when successive data points are strongly correlated. What practical diagnostic(s) would you use instead?
21. You are assigned to model a freeway work zone in a microscopic simulator (e.g. VISSIM). Draft a CVV plan including calibration targets/sequence, verification checks, validation at micro and macro levels, replication strategy, documentation, and accreditation decision questions.
22. Overfitting prevention strategy: Propose a method to prevent overfitting during calibration. Include (i) how to split data for calibration vs validation, (ii) how to test generalizability, and (iii) a stopping rule or constraint that limits over-tuning.

References

Annino, J. S. & Russell, E. C. (1979). Ten most frequency causes of simulation analysis failure-and how to avoid them. *Simulation*, *32*(6).

Balci, O. (1998). Verification, validation, and testing. In Banks, J. (ed.), *Handbook of Simulation: Principles, Methodology, Advances, Applications, and Practice*. Wiley, New York, Co-published by Engineering & Management Press.

Banks, J., Carson, J. S., Nelson, B. L., & Nicol, D. M. (2001). *Discrete-Event System Simulation* (3rd ed.). Prentice Hall, New Jersey.

Barcelo, J. & Casas, J. (2006). Stochastic Heuristic Dynamic Assignment Based on AIMSUN Microscopic Simulator, TRR No 1964, pp. 70–79.

Barceló, J. (ed.) (2010). *Fundamentals of Traffic Simulation*. Springer, New York, 2010.

Benekohal, R. F. (1986). Development and Validation of a Car Following Model for Simulation of Traffic Flow and Traffic Waves Studies. Ph.D. Dissertation, The Ohio State University, Columbus, Ohio.

Benekohal, R. F. (1991). Procedure for Validation of Microscopic Traffic Flow Simulation Models, TRR 1320, TRB, pp. 190–202, National Research Council, Washington, DC.

Benekohal, R. F. & Treiterer, J. (1989). A Car Following Model For Simulation of Traffic Flow in Normal and Stop-and-Go Conditions, TRR 1194, TRB, National Research Council, Washington DC.

Brockfeld, E., Kühne, R. D., & Wagner, P. (2004). Calibration and validation of microscopic traffic flow models. *Journal of Transportation Research Record, 1876*, 62–70. TRB, National Research Council, Washington, DC.

Casas, J., Ferrer, J. L., Garcia, D., Perarnau, J., & Torday, A. (2010). Chapter 5: Traffic simulation with Aimsun. In Barcelo, J. (ed.), *Fundamental of Traffic Simulation*. pp. 173–233, Springer, New York.

Chiu, Y., Bottom, J., Mahut, M., Paz, A., Balakrishna, R., Waller, T., & Hicks, J. (2011). Dynamic Traffic Assignment: A Primer. Published as Transportation Research Circular, TRB. DOI: 10.17226/22872.

Dowling, R., Skabardonis, A., Halkias, J., McHale, G., & Zammit, G. (2004). Guidelines for Calibration of Microsimulation Models: Framework and Applications. TRR, No. 1876, pp. 1–9, National Academies, Washington, DC.

Draper, N. & Smith, H. (1981). *Applied Regression Analysis* (2nd ed.). John Wiley & Son, Inc., New York.

Emeshoff, J. R. & Sisson, R. L. (1970). *Design and Use of Computer Simulation Models*. McMillan Co., New York.

Federal Highway Administration. (2004). Traffic Analysis Toolbox Volume III: Guidelines for Applying Traffic Microsimulation Modeling Software. FHWA-HRT-04-040.

Federal Highway Administration. (2007). Traffic Analysis Toolbox Volume IV: Guidelines for Applying CORSIM Microsimulation Modeling Software. FHWA-HOP-07-079.

Fellendorf, M. & Vortisch, P. (2010). Chapter 2: Microscopic traffic flow simulator VISSIM. In Barcelo, J. (ed.), *Fundamental of Traffic Simulation.* pp. 63–129, Springer, New York.

Fishman, G. S. (1967). Problems in the statistical analysis of simulation experiments: The comparison of means and the length of sample records. *Communications of the ACM, 10*(2), 94–99.

Fishman, G. S. & Kiviat, P. J. (1968). The statistics of discretion-even simulation. *SCI Simulation*, 10.

Gafarian, A. V. & Walsh, J. D. (1970). Methods for statistical validation of a simulation model for freeway traffic near an on-ramp. *Transportation Research, 4*, 379–324.

Gazis, D. C., Herman, R., & Rothery, R. W. (1961). Non-linear follow-the-leader models of traffic flow. *Operational Research, 9*, 545–567.

Hollander, Y. & Liu, R. (2008). The principles of calibrating traffic microsimulation models. *Transportation, 35*, 347–362. DOI: 10.1007/s11116-007-9156-2.

Hourdakis, J., Michalopoulos, P. G., & Kottommannil, J. (2003). Practical procedure for calibrating microscopic traffic simulation models. *Transportation Research Record, 1852*, 130–139.

Jehn, N. L. & Turochy, E. R. (2019). Calibration of Vissim models for rural freeway lane closures: Novel approach to the modification of key parameters. *Transportation Research Record, 2673*(5), 574–583.

Kan, X. D., Ramezani, H., & Benekohal, R. F. (2014). Calibration of VISSIM for freeway work zones with time-varying capacity. In *Presented at 93rd Annual Meeting of the Transportation Research Board*, Washington, DC.

Kleijnen, J. P. C. (1975a). Antithetic variates, common random numbers and optimal computer time allocation in simulation. *Management Science, 21*(10).

Kleijnen, J. P. C. (1975b). Statistical design and analysis of simulation experiments. *Informatie, 17*(10), 531–535. The Netherlands.

Kleijnen, J. P. C. (1976). Comparing means and variances of two simulations. *SCI Simulation, 26*(3).

Kleijnen, J. P. C. (1977). Design and analysis of simulations: Practical statistical techniques. *SCI Simulation, 28*(3).

Kleijnen, J. P. C. (1982). Experimentation with models: Statistical design and analysis techniques. In Cellier, F. E. (ed.), *Progress in Modeling and Simulation.* Academic Press: New York.

Kovvali, V. G., Alexiadis, V., & Zhang, L. (2007). Video-based vehicle trajectory data collection. In *TRB 86th Annual Meeting Compendium of Papers CD*, Washington, DC.

Krauss, S. (1998). Microscopic Modeling of Traffic Flow: Investigation of Collision Free Vehicle Dynamics. Doctoral Degree from the Faculty of Mathematics and Natural Sciences at the University of Cologne, Germany.

Law, A. M. & Kelton, W. D. (2000). *Simulation Modeling and Analysis* (3rd ed.). McGraw-Hill Inc., New York.

Mahut, M. & Florian, M. (2010). Chapter 9: Traffic simulation with Dynameq. In Barcelo, J. (ed.), *Fundamental of Traffic Simulation*, Springer, New York, pp. 323–361.

Menneni, S., Sun, C., & Vortisch, P. (2008). Microsimulation calibration using speed-flow relationships. *Journal of the Transportation Research Board, 2088*(1), 1–9.

Menneni, S., Sun, C. C., & Vortisch, P. (2009). Integrated microscopic and macroscopic calibration for psychophysical car-following models. In *TRB 88th Annual Meeting Compendium of Papers DVD*, Washington, DC.

Mihram, G. A. (1972). *Simulation Statistical Foundations and Methodology*. Academic Press, New York.

Naylor, T. H., Wertz, K., & Wonnacott, T. H. (1967). Methods for analyzing data from computer simulation experiments. *Communication of ACM, 10*(11).

Petty, M. D. (2010). Chapter 10: Verification, validation, and accreditation. In Sokolowski, J. A. & Banks, C. M. (eds.), *Modeling and Simulation Fundamentals*, Wiley & Sons, New Jersey, pp. 325–373.

Rakha, H. & Wang, W. (2009). Procedure for Calibrating Gipps Car-Following Model, TRR, No. 2124, pp. 113–124, National Academies, Washington, DC.

Sargent, R. S. (1982). Verification and validation of simulation models. In Cellier, F. E. (ed.), *Progress in Modeling and Simulation*. Academic Press: New York.

Skyes, P. (2010). Chapter 4: Traffic simulation with paramics. In Barcelo, J. (ed)., *Fundamental of Traffic Simulation*. pp. 131–171, Springer, New York.

Sokolowski, J. A. & Banks, C. M. (eds.). (2010). *Modeling and Simulation Fundamentals: Theoretical Underpinnings and Practical Domains*. Wiley & Sons, New Jersey.

Toledo, T., Koutsopoulos, H. N., Davol, A., Ben-Akiva, M. E., Burghout, W., Andreasson, I., Johansson, T., & Lundin, C. (2003). Calibration and validation of microscopic traffic simulation tools – Stockholm case study. *Transportation Research Record, 1831*, 65–75.

Treiterer, J. (1975). Investigation of Traffic Dynamics by Aerial Photogrammetry Techniques. Transportation Research Center Department of Civil Engineering, The Ohio State University, Final Report EES 278, February 1975.

Wisconsin DOT, Department of Transportation. (2019). Chapter 16: Traffic engineering, operations & safety manual. In *Section 20: Traffic analysis and modeling. Microscopic Simulation Traffic Analysis*, September 2019.

Zhao, L., Rilett, L. R., & Hague, M. S. (2022). Calibration and validation methodology for simulation models of intelligent work zones. *Transportation Research Record, 2676*(5), 500–513.

Chapter 7

Statistical Validation Methods

There are several methods (techniques) for validating simulation results. Typically, they involve comparing simulation results to reliable reference data (benchmark data, field data, etc.) to assess how similar they are and attempt to gain confidence that the simulation results are valid. The benchmark data often are field data or other data sources that can reliably represent the system performance. The validation methods (approaches) may be divided into two broad categories: (a) visual methods and (b) statistical methods.

7.1 Visual Methods

In the visual methods, the simulation outcomes are visually compared to field data. The visual comparison may look at the mean, mode, variance, certain percentile values, or may even look at the distribution of the results without conducting any statistical analysis. One may look at the histogram of results, profile of outcomes over time and/or space, or fluctuation of the results as systems input values or network operating conditions change. The simulation results may be compared, without statistical analysis, to an expected outcome the user has in mind, certain threshold values, or the highest or lowest values over time or space (upper or lower limit). The visual method by its nature is a subjective approach because there is no statistical analysis conducted. It is easy to do, but the conclusions in favor of or against the simulation may not be very convincing, as it depends largely on the judgment of the analyst. It is a useful approach

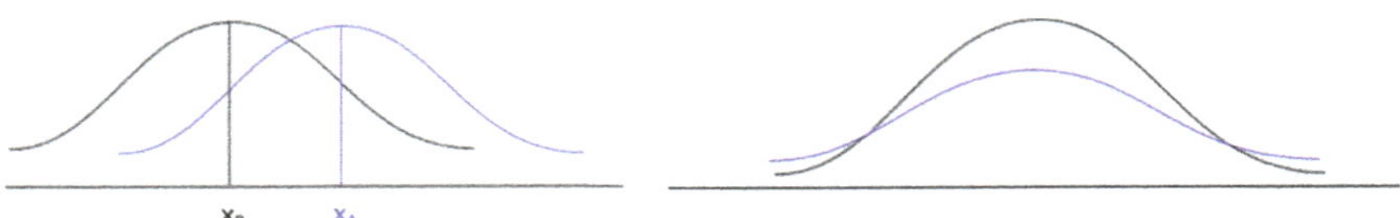

Figure 7.1. Comparing means and distribution of simulation results to benchmark data.

to assess how close or far the simulation results are from real-world systems.

An example of the visual method is illustrated in Figure 7.1. On the left sketch, the mean of the distribution of simulation results (X_B) seems to be close to the mean of benchmark data (X_A). On the right sketch, the shape (distribution of data) for data representing the simulation seems to be very similar to the shape of the benchmark data. One person may judge the two means to be close enough to each other ($X_B \approx X_A$), and another person may say they are far because the judgments are based on the visual assessment and thus subjective (and not objective). Similarly, one person may say the distributions (shapes) are very similar, and another person may challenge that notion. The shortcoming of the visual method is its subjectivity. To avoid subjectivity in our assessments, we need objective approaches that rely on statistical analyses.

7.2 Statistical Methods

In contrast to the visual method, the statistical methods rely on conducting statistical analyses to determine the similarity or dissimilarity of the simulation results compared to the field data (benchmark data). Statistical methods, when they are done properly, can objectively assess the similarity (or lack of) of the simulation results to the field data. There are many statistical methods one may use to validate simulation results. Some of the common statistical validation methods are

- regression analysis;
- time series analysis;
- categorial analysis;
- comparing mean values (*z*-test, *t*-test);
- comparing the pairs of observations (paired sample *t*-test);
- comparing goodness of fit of data (chi-square or K–S test);
- comparing the variance of distributions (*F*-test).

There are different statistical methods that are used in model validation. What methods should be used depends on the skill levels of the analyst and the nature of the data available. This chapter presents the widely used statistical analysis techniques that can be used in the validation of simulation models. The statistical tests presented in this chapter are the z-test, t-test, paired sample t-test, F-test, chi-square test, and Kolmogorov–Smirnov (K–S) test for goodness of fit. For further information, consult modeling and simulation books such as Banks *et al.* (2001), Fishman (1978), Law and Kelton (2000), Mihram (1972), Pooch and Wall (1992), Pritsker and Pegden (1979), Seila *et al.* (2003), Sokolowski *et al.* (2010), and Ross (2002), as well as statistics books such as Box *et al.* (2005), Draper and Smith (1981), Hogg *et al.* (2021), Kirk (2013), Montgomery (2019), Ott and Longnecker (2015), and Snedecor and Cochran (1989).

7.2.1 *Selecting the Test (z-Test, t-Test, or Paired Sample t-Test)*

The type of data available and its size, to a large degree, determine which test should be conducted. For comparing the average performance measure (e.g. speed, travel time, and headways) from a simulation model to field data (or benchmark data), one may use classical tests such as the z-test, t-test, and paired sample t-test. To compare the variance of the performance measure to the variance of field data, one may use the F-test. To compare the distribution of the performance measure from the simulation model to the field data, one may use the chi-square test or the Kolmogorov–Smirnov (K–S) test.

The following example illustrates when to use the z-test, t-test, or paired sample t-test.

7.2.2 *Pairable Data*

Consider a situation where you have field data for n conditions (F_1, F_2, F_3, $\ldots F_n$). For each of these conditions, you also have corresponding results from simulation models (S_1, S_2, S_3, $\ldots S_n$), as shown in Table 7.1. Here, field data for condition 1 ($F1$) has a corresponding result from the simulation model ($S1$). So, the data are pairable.

For the situation described above (pairable condition), you should do a paired sample t-test (described later) because for each field condition, you have a comparable simulation representing that condition. The objective of pairing is to increase the accuracy of comparison of the two sets (Snedecor & Cochran, 1989).

Table 7.1. Paired data and paired *t*-test.

Pairable Condition	Simulation Results	Field Data	Paired Differences
1	S_1	F_1	$D_1 = S_1 - F_1$
2	S_2	F_2	$D_2 = S_2 - F_2$
3	S_3	F_3	$D_3 = S_3 - F_3$
⋮	⋮	⋮	⋮
n	S_n	F_n	$D_n = S_n - F_n$
Column Average	$\overline{S}$	$\overline{F}$	$\frac{\Sigma_1^n D_i}{n}$
Variance	Var$S$	Var$F$	Var$D$

Table 7.2. Non-pairable data and *t*-test.

	Simulation Results	Field Data
Non-Pairable Condition	S_1	—
	S_2	—
	S_3	—
	S_4	F_1
	S_5	F_2
	S_6	F_3
	⋮	⋮
	S_m	F_n
Column Average	$\overline{S}$	$\overline{F}$
Variance	Var$S$	Var$F$

7.2.3 *Non-Pairable Data*

When F_1 is not comparable to S_1, S_2 is not comparable to S_2, and F_n is not comparable to S_n, as shown in Table 7.2, this indicates the data are not pairable. For non-pairable conditions (Table 7.2 example) with two independent samples with means $\overline{S}$ and $\overline{F}$, you may do either the *z*-test or *t*-test, depending on the assumption about the variances (known or unknown) and sample sizes. It is assumed that $\overline{S}$ and $\overline{F}$ are normally distributed and independent. If this assumption is correct, their difference $(\overline{S} - \overline{F})$ is also normally distributed (Snedecor & Cochran, 1989).

7.2.4 *When to Use z- and t-Test*

For not-pairable conditions, use *z*-test or *t*-test (some refer to *t*-test as Student's *t*-test; they are synonyms). Which test should I use *z*-test or *t*-test? If the variances are known and assumed to be equal, one can do a *z*-test. If the variances are unknown and must be computed from the sample data and they are assumed to be equal, do a *t*-test. The sample size from the field may/may not be equal to the sample size from the simulation model. In the example, there are "*n*" results from field data and "*m*" results from the simulation model. If both *m* and *n* are larger than or equal to 30, do a *z*-test. Of course, one can do a *t*-test in this condition as well, and the results will not be a lot different from a *z*-test since the sample size is larger than 30.

What Statistical Test Should be Used

- ***When data is not pairable***

  If $m \geq 30$ and $n \geq 30$, variances are known ***Use z-test***

  If m or n < 30, variances are unknown,

  but computed from sample data ***Use t-test***

- ***When data is pairable*** ***Use paired sample t-test***

When the sample size is larger than 30, there is a small difference between the *z* and *t* distributions, and that difference is mainly near the tails of the curve (refer to Law & Kelton, 2000). So, one can do *t*-test when it is justified to do a *z*-test. If *m* or *n* is smaller than 30, do *t*-test. Also, when the sample size from the field, *n*, is not equal to the sample size from the simulation, *m*, then use *t*-test instead of paired sample *t*-test.

For further discussion about the conditions for pairing and the advantages and disadvantages of pairing, refer to Chapter 6 of Statistical Methods by Snedecor and Cochran (1989). They examine the formula for the variance of the differences and role of covariance in reducing the variance of the difference.

7.2.5 *t*-Test

First, compute the average value for the field data ($\overline{F}$) and the average value for the simulation outcomes ($\overline{S}$):

$$\overline{F} = \frac{\sum_{1}^{n}(F_i)}{n}$$

$$\overline{S} = \frac{\sum_{1}^{m}(S_i)}{m}$$

Then, find the variance of each set:

$$\text{Var}F = \frac{\sum_{1}^{n}\left(F_i - \overline{F}\right)^2}{n-1}$$

$$\text{Var}S = \frac{\sum_{1}^{m}\left(S_i - \overline{S}\right)^2}{m-1}$$

where F_i is the ith observation in the field, $\overline{F}$ is the average of field observations, S_i is the ith observation in the simulation, $\overline{S}$ is the average of simulation outcomes, Var$F$ is the variance of field data, S_F is the standard deviation of field observations, Var$S$ is the variance of simulation results, Var$P$ is the pooled variance, S_P is the standard deviation of pooled variance, S_S is the standard deviation of simulation results, n is the number of field observations, and m is the number of simulation results.

Then, compare $\overline{F}$ with $\overline{S}$ by finding the difference between the two average values (D):

$$D = \overline{S} - \overline{F}$$

Here, the null hypothesis (H_0) is that D is not significantly different than zero:

$$H_0\text{: } D = 0$$

The alternative hypothesis (H_a) is that it is significantly different than zero:

$$H_a\text{: } D \neq 0$$

Table 7.3. Standard error (SE) formula for different tests.

Test Type	Formula for SE
t-test when variances are equal	$S_P \sqrt{\frac{1}{n}+\frac{1}{m}}$
t-test when variances and sample sizes are equal	$S_P \sqrt{\frac{2}{n}}$
t-test when variances are unequal (Welch)	$\sqrt{\frac{\text{var}F}{n}+\frac{\text{Var}S}{m}}$
Paired *t*-test	$S_d / \sqrt{n}$

Compute t (or z) value by dividing the difference (D) by the standard error (SE):

$$t\ (\text{or } Z) = \frac{\overline{S} - \overline{F}}{SE}$$

The formula for SE are given in Table 7.3. If n or m is less than 30, do a t-test, but if both n and m are 30 or larger, do a z-test.

7.2.6 *What Variance Should Be Used*

The next step is to compute the variance. What equation should be used to compute? There are two formulas to compute the variance, depending on the assumptions one makes about the equality of variances.

7.2.7 *Equal Variances*

If the variance of the simulation results is equal to the variance of the field data, compute a pooled variance (Var$P$) from the following equation:

$$\text{Var}P = \frac{(n-1)VarF + (m-1)VarS}{n+m-2}$$

The pooled variance (Var$P$) equation simplifies to the following equation when the sample sizes are the same ($m = n$):

$$\text{Var}P = \frac{(\text{Var}F + \text{Var}S)}{2}$$

7.2.8 *Unequal Variances*

If the variance of the simulation results is not equal to the variance of the field data, do a two-sample t-test with unequal variances, which is called Welch's t-test. Standard error for unequal variance (Weltch test) is

$$\text{SE} = \sqrt{\frac{\text{var}F}{n} + \frac{\text{Var}S}{m}}$$

7.2.8.1 *Example for t-Test*

Let's assume the average speed of vehicles on a link is measured in the field 10 different times (F_1, F_2, F_3, …, F_{10}), as shown in Table 7.4. Then, average speeds on the same link are obtained from a simulation. Let's assume 10 simulation results are obtained, even though it does not need to be equal to the number of data points from field. The simulation programs are replicated enough times, and the average values of the replications is given in Table 7.4 as S_1, S_2, S_3,….S_{10}. Note that S_1 does NOT corresponds to F_1 and S_2 does NOT correspond to F_2, and so on. So, we cannot pair them.

Now, compute t value and compare it to the threshold value computed from the t-distribution (called t table):

$$t = \frac{\overline{S} - \overline{F}}{SE}$$

$$\text{SE} = S_P \sqrt{\frac{1}{n} + \frac{1}{m}}$$

$$S_p = \sqrt{\text{Var}P} = \sqrt{\frac{(n-1)\text{Var}F + (m-1)\text{Var}S}{n+m-2}}$$

Table 7.4. Example problem for *t*-test.

Number of Observations	Field Data	Number of Observations	Simulation Results
10	36.1	10	34.9
	30.8		32.1
	30.6		32.3
	32.3		31.4
	33.5		32.8
	31.4		33.2
	30.2		31.7
	33.6		35.1
	33.9		32.5
	34.7		32.6
Average	32.71	Average	32.86
Variance	3.841	Variance	1.5404

$$S_p = \sqrt{\frac{(10-1)3.841+(10-1)1.5404}{10+10-2}} = 1.6403$$

$$\text{SE} = 1.6403\left(\sqrt{\frac{1}{10}+\frac{1}{10}}\right) = 0.7336$$

$$t = \frac{32.71-32.86}{0.7336} = -0.204$$

The t table for 18 degrees of freedom ($10 + 10 - 2 = 18$) for a two-sided test with an α value of 5% is 2.101. Since the computed t value is less than the t table ($0.204 < 2.101$), we do not reject the hypothesis.

7.2.9 *Paired Sample t-Test*

If F_1 is comparable to S_1, S_2 is comparable to F_2, and F_n is comparable to S_n, then the data are pairable. For paired data, conduct a paired sample t-test by finding $D_i = S_i - F_i$. Then, treat these differences as the data points to be analyzed and find their average and variance. It should be noted that paired sample *t*-test is done when we have paired data, meaning F_1 can be compared to S_1, F_2 can be compared to S_2, and so on.

For the paired sample *t*-test, use $D_i = S_i - F_i$. Find $\bar{D}$ which is the average of the differences. Also, compute the variance of them (Var*D*) from the following equations:

$$\bar{D} = \frac{\sum_1^n (D_i)}{n}$$

$$\text{Var}D = \frac{\sum_1^n \left(D_i - \bar{D}\right)^2}{n-1}$$

where F_i is the *i*th observation in the field, S_i is the *i*th observation in the simulation, D_i is the difference of the paired observations, $D_i = S_i - F_i$, $\bar{D}$ is the average value of the difference (D_i), Var*D* is the variance of the differences, and *n* is the number of field observations (*n* is also the number of simulation observations).

Then, perform a test to see if $\bar{D}$ is significantly different than zero. If *n* is smaller than 30, do a *t*-test; otherwise, do a *z*-test. Here, the null hypothesis (H_0) is that $\bar{D}$ is not significantly different than zero, and the alternative hypothesis (H_a) is that it is significantly different than zero. Note that the degree of freedom (df) for the paired sample *t*-test is (n–1):

$$H_0\text{: } \bar{D} = 0$$

$$H_a\text{: } \bar{D} \neq 0$$

To conduct a *t*-test or a *z*-test, the confidence level to be used must be decided. The higher the confidence level, the larger the difference between the simulation and field data must be. Often, the confidence level is 90% is used, and it is an acceptable level for most comparisons. However, one may set a confidence level to 95% to have higher confidence in the results of the comparison. The higher confidence level is often used when the sample size is large.

7.2.9.1 *Example for Paired Sample t-Test*

Average speed of vehicles at a certain location on a link is measured in the field for 10 different conditions (F_1, F_2, F_3, …, F_{10}), as shown in Table 7.5. Then, 10 simulation programs are made each corresponding to one of the

Table 7.5. Example for paired data and paired t-test.

Conditions	Field Data	Simulation Results	D (Field Simulation)	Paired t-Test Calculations $D - \bar{D}$	$(D - \bar{D})^2$
1	36.1	34.9	1.2	1.35	1.8225
2	30.8	32.1	−1.3	−1.15	1.3225
3	30.6	32.3	−1.7	−1.55	2.4025
4	32.3	31.4	0.9	1.05	1.1025
5	33.5	32.8	0.7	0.85	0.7225
6	31.4	33.2	−1.8	−1.65	2.7225
7	30.2	31.7	−1.5	−1.35	1.8225
8	33.6	35.1	−1.5	−1.35	1.8225
9	33.9	32.5	1.4	1.55	2.4025
10	34.7	32.6	2.1	2.25	5.0625
Average	—	—	−0.15	Sum	21.205
Variance	—	—	2.356111*	Sum/(9)	2.356111

Note: *Variance function in the Excel spreadsheet computed this value.

field conditions. For each condition, the simulation programs are replicated enough times, and the average values of the replications are given in Table 7.5 as $S_1, S_2, S_3, \ldots S_{10}$. Thus, S_1 corresponds to F_1, S_2 corresponds to F_2, and so on.

Then, compute t value and compare it to the threshold value computed from the t-distribution (called t table):

$$t = \frac{\bar{D}}{S_d / \sqrt{n}} = \frac{-0.15}{\sqrt{2.356111} / \sqrt{10}} = -0.30902$$

The t table for 9 degrees of freedom ($10 - 1 = 9$) for a two-sided test with α value of 5% is 2.262. Since the computed t value is less than the t table ($0.3090 < 2.262$), we do not reject the hypothesis. The hypothesis was that the simulation results and field data are not significantly different. Thus, the simulation results are not significantly different than the field data.

Comparing the variances shows that the variance of the paired t-test is slightly larger than the variance of the t-test ($2.35611 > 2.31985$). This difference should be considered with the df for each case in finding the

t values from the table. In a paired t-test, the df is 9, while in an independent sample t-test, the df is 18. Thus. The t values from tables would be different. Assuming the simulation results are independent of the field data and thus could not be paired yielded lower variance of difference than paining them. This indicates that the covariance term should be ignored, as it is done by assuming independence. For further discussion, refer to Snedecor and Cochran (1989).

7.2.10 *F-Test for Equality of Two Variances*

To compare the variance of simulation results to the variance of field data, one can run an F-test. The F-test is conducted to test if variances are equal. An F-test is needed to establish that the data in both cases come from a similar distribution. If you establish that the average of simulation outcomes is not significantly different than the average of the field data points, and establish that variances are also the same, then you can claim the two distributions are not significantly different. To conduct an F-test for comparison of two variances, compute F as the ratio of the two variances (larger variance over smaller variance):

$$F = \text{Var}F/\text{Var}S$$

The computed F value is always positive since the numerator and denominator are positive. If they are equal, then F is one. If not equal, then the F value will be greater than one. This calculated F value will be compared to a value from the F distribution for a given degree of freedom and confidence level. To find df for the F test for two variance comparison, first find df for the numerator, which is $m - 1$, and then find df for the denominator, which is $n - 1$.

The computed F value will be compared to an F value obtained from an F-distribution function (table) for a given significance level (α). The null hypothesis is that the variances are equal, and the alternative hypothesis is that they are not equal:

$$H_0\text{: Var}S = \text{Var}F$$

$$Ha\text{: Var}S \neq \text{Var}F$$

F-test can be conducted as a one-sided or a two-sided test depending on the way the hypothesis is set up. It is a one-sided F-test if the hypothesis states that one variance is greater than the other. It is a two-sided F-test if the hypothesis states that the variances are different without saying which one is greater than the other. In the above setting, it will be a two-sided test because it is not testing which one of the variances is greater than the other one.

7.2.10.1 *Example for F-Test*

The example that was used for t-test will be used to do F-test. In that example, the variance of field speed data was 3.841, and the variance for simulation speed results was 1.5404. Compute F,

$$F = \text{Var}F/\text{Var}S = 3.841/1.5404 = 2.494$$

The value of the F-distribution function (F-Table) for df of 9 for the numerator and df of 9 for the denominator with an α value of 5% is 3.18. Since the computed F value is less than the F table ($2.494 < 3.18$), we do not reject the hypothesis. The hypothesis is that the variance of the simulation results is not significantly different than the variance of field data.

7.2.11 *Goodness of Fit Tests*

In addition to the above test, two other statistical tests (chi-square and K–S) are often used to compare the goodness of fit of data coming from a simulation model to the data from field. The chi-square goodness of fit test compares the distributions of the simulation and filed data. While chi-square test uses probability density function (PDF), the Kolmogorov–Smirnov (K–S) test uses the cumulative density functions (CDF) of the two datasets to examine how large the maximum difference between the two cumulative curves.

7.2.12 *Chi-Square Goodness of Fit Test*

Sometimes, the random variable of interest (such as the outcome of the simulation) may follow a particular distribution. This distribution may be compared to a mathematical distribution as an observed distribution. To make such a comparison, a chi-square goodness of fit test may be conducted to see whether the distribution is significantly different than the mathematical or the observed one. This test makes a comparison between the "actual" and "expected" number of observations. Here, the "actual" is the simulation results and the "expected" is the numbers from a mathematical distribution or from observed field data. So, chi-square (χ^2) is computed as

$$\chi^2 = \sum_{1}^{r} \frac{(O_i - E_i)^2}{E_i}$$

where χ^2 is the computed chi-square, O_i is the observed frequency of simulation outcomes in the ith interval, E_i is the expected frequency from a mathematical distribution of field data in the ith interval, and r is the number of intervals (categories or classes).

The number of categories (r) should be sufficient to give an accurate picture of the distribution. For example, if it is a normal distribution, then the number of intervals should be sufficient to give a plot that looks like a normal distribution. This may need about 10 intervals. The number of intervals depends on the width of the interval selected. The number of intervals selected is somewhat arbitrary (Banks *et al.*, 2001, p. 348; Pooch & Wall, 1992); however, less than three to five may distort the results (Pooch & Wall, 1992). Changing the number of intervals affects the calculated chi-square (Banks *et al.*, 2001, p. 348). Also, it changes the theoretical chi-square value (table value) because of the change in df. A minimum number of expected frequencies for each interval is needed, and some have used 3, 4, and 5 (Banks *et al.*, 2001, p. 344). Often, five is mentioned in textbooks. For continuous data, the number of intervals suggested by Banks *et al.* (2001, p. 344) is 5–10 if the sample size is 50, and 10–20 if sample size is 100.

The df for this test is ($r - p - 1$). Here, p is the number of parameters estimated for mathematical distribution. For example, for a normal distribution, p is 2 because the mean and variance are the two estimated parameters. For example, if 12 intervals ($r = 12$) are used for a normal distribution ($p = 2$), then df $= 12 - 2 - 1 = 9$. For the Poisson distribution, p is one since we estimate only one parameter (mean and variance are equal for Poisson distribution). The computed chi-square is compared to the chi-square value from a chi-square function with for a given α and the df. If the computed χ^2 is greater than or equal to the χ^2 from the distribution (table), reject the hypothesis that the observed and actual distribution are the same, implying that the simulation outcomes are not from the same distribution as the field data:

$$\chi^2 \geq \chi^2_{(r-p-1,\alpha)}$$

A hypothesis may be accepted when data are grouped in one way and may be rejected when they are grouped in another way (Banks *et al.*, 2001, p. 348) since the chi-square test result is affected by interval width and number of intervals. Another test for the goodness of fit examination that does not have this issue is the K–S test. For comparison of the chi-square test to the K–S test, you may look at Law & Kelton (2000).

7.2.12.1 *Example for Chi-Square Goodness of Fit Test*

A simulation produces 120 observed values grouped into 8 intervals. Field data provide expected frequencies for the same intervals.

Interval	**Observed O_i (Simulation)**	**Expected E_i (Field)**	$(O_i-E_i)^2$	$((O_i-E_i)^2)/E_i$
0–1	6	8	4	0.5000
1–2	14	12	4	0.3333
2–3	22	20	4	0.2000
3–4	28	30	4	0.1333
4–5	20	18	4	0.2222
5–6	16	14	4	0.2857
6–7	10	12	4	0.3333
7–8	4	6	4	0.6667

Notes: Chi-square test statistic is $\chi^2 = \Sigma\,(O_i - E_i)^2/E_i = 2.6746$.

Degrees of freedom assuming no distribution parameters are estimated, df = 8 − 1 = 7.

For df = 7 at $\alpha = 0.05$, critical value $\chi^2 = 14.067140$ and p-value = 0.913386.

Therefore, do not reject the hypothesis H_0, implying that the simulation results are not significantly different than field data.

7.2.13 *Kolmogorov–Smirnov (K–S) Test*

Another test for assessing if two distributions are similar is the Kolmogorov–Smirnov (K–S) test. In the K–S test, the cumulative distribution function (CDF) of the simulation outcomes is plotted and compared to a theoretical CDF (or to a CDF for field data). The vertical differences (D) between the two cumulative distributions are found, and the greatest difference is used to assess if the distributions are similar. The K–S test is useful when sample sizes are small and no parameters are estimated using the data (Banks *et al.*, 2001, p. 348). When parameters are estimated from the data, the K–S test can be used in several cases, but different tables of critical values are required (Banks *et al.*, 2001, p. 348). The K–S test is a non-parametric test, meaning it does not make any assumption about underlaying distribution of the data. Steps in conducting the K–S test are as follows:

1. Compute the CDF for the simulation results, $F_s(x)$.
2. Compute the theoretical CDF (if comparing it to field data, find CDF for the field data), $F_t(x)$.
3. Compute the difference between the two CDF at x_i, $D_i = F_s(x_i) - F_t(x_i)$. Find the absolute value of D_i, $|D_i|$.
4. Find the largest value of the absolute differences, $D = \max |F_s(x) - F_t(x)|$.
5. Find the critical value for K–S from a table (formula). The table K–S value depends of sample size and α values.
6. If the calculated D is greater than the table value, reject the hypothesis.

For more information on the chi-square test and KS test, you may refer to these references: Simulation by Ross (2002), Pooch and Wall (1992), and Banks *et al.* (2001).

7.2.13.1 *Example for K–S Test Goodness of Fit Test*

At selected points x, the empirical CDF of simulation outcomes $F_s(x)$ and the empirical CDF of field data $F_t(x)$ are given in the following.

x	$F_s(x)$ Simulation	$F_t(x)$ Field	$\|F_s(x) - F_t(x)\|$
10	0.08	0.05	0.03
20	0.22	0.18	0.04
30	0.44	0.40	0.04
40	0.65	0.70	0.05
50	0.84	0.88	0.04
60	0.93	0.95	0.02

Sample size $n = 75$. Use $D = \max|F_s(x) - F_t(x)|$ and approximate critical value $D\_crit = 1.36/\sqrt{n}$ for $\alpha = 0.05$.

Compute $|F_s(x) - F_t(x)|$ at each x. Determine Dmax. Dmax is 0.05.

Compute $D\_crit = 1.36/8.6603 = 0.15704$.

Therefore, do not reject the null hypothesis (H_0). H_0: The two distributions are not significantly different at $\alpha = 0.05$.

Exercises

Part 1. Questions

1. Explain the difference between visual validation methods and statistical validation methods. Discuss the advantages and disadvantages of each approach.
2. Why is the visual validation method considered subjective? Provide two examples where visual validation may lead to incorrect conclusions.
3. Define each test and explain its purpose in simulation validation: (a) z-test, (b) two-sample t-test, (c) paired sample t-test, (d) F-test, (e) chi-square goodness-of-fit test, and (f) Kolmogorov–Smirnov (K–S) test.
4. Under what conditions should (a) a z-test be used instead of a t-test, (b) a paired t-test be used instead of an independent two-sample t-test, and (c) Welch's t-test be used?
5. Explain why pairing observations may reduce the variance of the difference between two samples.
6. What is pooled variance, and when do we use it?
7. What is the main difference between standard deviation and standard error?
8. State key assumptions for (a) two-sample t-test, (b) paired t-test, (c) F-test, and (d) chi-square test.
9. Why does the chi-square test depend on the number of intervals selected? Why does the K–S test not suffer from this issue?
10. Both the chi-square test and K–S test are used for checking the goodness of fit of a distribution. What should be considered in choosing one over the other?

Part 2. Problems

Problem 1. Two-sample t-test (equal variances).

Field travel time (minutes), $n = 12$: 12.1, 13.5, 11.8, 12.9, 13.2, 12.4, 12.8, 13.0, 12.6, 13.1, 12.2, 13.4.

Simulation travel time (minutes), $m = 12$ (independent runs): 11.9, 12.7, 12.4, 12.3, 12.5, 12.8, 12.6, 12.9, 12.2, 12.4, 12.7, 12.5:

a. Compute the sample mean and sample variance of both samples.
b. Compute the pooled variance and pooled standard deviation.
c. Conduct a two-sided two-sample t-test at $\alpha = 0.05$ (assume equal variances). State hypotheses, test statistic, critical value (or p-value), and conclusion.
d. Based on the test, is the simulation statistically consistent with field data with respect to the mean travel time?

Problem 2. Welch's t-test (unequal variances).

Field delay (seconds), $n = 11$: 42, 45, 39, 41, 44, 40, 43, 46, 38, 47, 41.

Simulation delay (seconds), $m = 15$: 38, 37, 41, 39, 40, 36, 42, 39, 38, 37, 40, 41, 36, 39, 38:

a. Compute the sample means and sample variances.
b. Conduct Welch's two-sample t-test at $\alpha = 0.05$, state hypotheses, test statistic, degrees of freedom, and conclusion.
c. Interpret the result for simulation model validity with respect to the mean delay.

Problem 3. Paired sample t-test.

Condition	Field (vehicles)	Simulation (vehicles)
1	15	14
2	18	17
3	14	16
4	20	19
5	17	15
6	16	18
7	19	18
8	13	14
9	21	20

Use paired differences D_i = Simulation$_i$ – Field$_i$:

a. Compute the paired differences, the mean difference, and the variance of differences.
b. Conduct a paired t-test at $\alpha = 0.05$. State hypotheses, test statistic, critical value (or p-value), and conclusion.
c. Interpret the result.

Problem 4. Choosing the correct test.
For each scenario, state which test should be used and justify briefly:

a. 75 simulation results compared to 83 field observations (independent).
b. 18 simulation results where each one corresponds to a specific field condition (pairable).
c. 23 simulation observations and 23 independent field observations, unknown variances.
d. Comparing the variability of simulation results vs field data.

Problem 5. F-test for equality of two variances.
Using the data from Problem 1:

a. State hypotheses for a two-sided F-test for equality of variances.
b. Compute the F statistic (use the larger variance in the numerator).
c. Find the critical value(s) at $\alpha = 0.05$ and state your conclusion.
d. Based on the result, comment on whether the equal-variance assumption is reasonable.

Problem 6. Chi-square goodness-of-fit.
Simulation headway data ($n = 90$) grouped into six intervals:

Interval	Observed O_i (Simulation)	Expected E_i (Field)
0–2	8	9
2–4	18	15
4–6	27	30
6–8	21	18
8–10	10	12
10–12	6	6

a. Compute the chi-square test statistic.
b. Determine degrees of freedom (assume no distribution parameters are estimated from these data).
c. At $\alpha = 0.05$, state whether to reject or fail to reject the hypothesis that the distributions are the same.

Problem 7. Kolmogorov–Smirnov (K–S) test.

x	$F_s(x)$ Simulation	$F_f(x)$ Field
1	0.10	0.15
2	0.25	0.30
3	0.50	0.55
4	0.75	0.70
5	0.90	0.88

Sample size $n = 40$. Use $D = \max|F_s(x) - F_f(x)|$. Use K–S critical value approximately $D\_crit = 1.36/\sqrt{n}$ for $\alpha = 0.05$:

a. Compute $|F_s(x) - F_f(x)|$ at each x and find D_max.
b. Compute D_crit and state your conclusion at $\alpha = 0.05.3$.

Part 3. Analytical and Discussion Problems

1. Explain why failing to reject the null hypothesis does not prove the simulation model is correct.
2. Discuss the role of confidence level (e.g. 90% vs 95%) in model validation. How does increasing confidence level relate to Type I and Type II errors?
3. A model passes a mean comparison test but fails a variance comparison test. Discuss possible reasons and implications.
4. Compare chi-square and K–S tests in terms of sensitivity, sample size requirements, assumptions, and strengths/weaknesses.
5. You have 53 field observations and 53 independent simulation replications. Describe a complete statistical validation procedure (mean comparison, variance comparison, and distribution comparison) and how you would interpret results.

References

Banks, J., Carson, J. S., Nelson, B. L., & Nicol, D. M. (2001). *Discrete-Event System Simulation* (3rd ed.). Prentice Hall, New Jersey.

Box, G. E. P., Hunter, J. S., & Hunter, W. G. (2005). *Statistics for Experimenters: Design, Innovation, and Discovery* (2nd ed.). Wiley-Interscience, New York, NY.

Draper, N. & Smith, H. (1981). *Applied Regression Analysis* (2nd ed.). John Wiley & Sons, Inc., New York.

Fishman, G. S. (1978). *Principles of Discrete Event Simulation.* Wiley, New York.

Hogg, R. V., Tanis, E. A., & Zimmerman, D. (2021). *Probability and Statistical Inference* (10th ed.). Pearson, New York, NY.

Kirk, R. E. (2013). *Experimental Design: Procedures for the Behavioral Sciences* (4th ed.). SAGE Publications, Inc., Thousand Oaks, CA.

Law, A. M. & Kelton, W. D. (2000). *Simulation Modeling and Analysis* (3rd ed.). McGraw-Hill Inc., New York.

Mihram, G. A. (1972). *Simulation Statistical Foundations and Methodology*. Academic Press, New York.

Montgomery, D. C. (2019). *Design and Analysis of Experiments* (10th ed.). Wiley, New York, NY.

Ott, R. L. & Longnecker, M. (2015). *An Introduction to Statistical Methods and Data Analysis* (7th ed.). Cengage Learning, Boston, MA.

Pooch, U. W. & Wall, J. A. (1992). *Discrete Event Simulation: A Practical Approach*. CRC Press, Boca Raton, FL.

Pritsker, A. A. B. & Pegden, C. D. (1979). *Introduction to Simulation and SLAM*. John Wiley & Sons, New York.

Ross, Sheldon M. (2002). *Simulation* (3rd ed.) Academic Press, San Diago.

Seila, A. F., Ceric, V., & Tadikamalla, P. R. (2003). *Applied Simulation Modeling*. Thomson, Brooks/Cole, Belmont, CA.

Snedecor, G. W. & Cochran, W. G. (1989). *Statistical Methods* (8th ed.). Iowa State University Press, Ames.

Sokolowski, J. A. & Banks, C. M. (eds.) (2010). *Modeling and Simulation Fundamentals: Theoretical Underpinnings and Practical Domains*. Wiley & Sons, New Jersey.

Chapter 8

Output Variability Analysis and Random Number Generation

Output analysis refers to examining the results obtained from the simulation model to assess how well it represents the real system. The simulation outputs may vary significantly due to the use of stochastic elements in defining the system variables (internal variability). Of course, the simulation output may change due to changes in input variables (external variability). One needs to understand the causes of the variability and how to deal with the effects of it. For more detailed discussion on output analysis, refer to references such as Banks *et al.* (2001), Fishman (1978), Law and Kelton (2000), Pooch and Wall (1992), Sokolowski *et al.* (2010). Stochasticity in the system variable can have many effects, and two important ones are the effect on confidence level and the effect on sample size. In this chapter, the sources of variability, the relationship between internal and external variability, how to deal with the variabilities, how to determine the length of the transient state, and random number generation issues are discussed.

8.1 Variability Sources

The user may get an incorrect impression that the system outputs are "constant" from one run to the next if the input variables are kept the same. The seemingly "constant" results should not be interpreted as "system has reached a steady state" condition, the "system providing consistent results", or the "lack of variability in the system outputs".

Rather, this should be considered as the output of one of the many runs of the same system with no changes in the input vairable.

There are two main sources of variabilities in system output that both need to be examined, and their impacts need to be determined before using the simulation results. The two sources of variability are

1. external variability sources;
2. internal variability sources.

8.1.1 *External Variability Sources*

Understanding the role and impact of external sources that cause variability in simulation output is relatively easy. External variability is due to changes in the input values that characterize geometry, traffic, roadway system, and traffic management decisions. For example, when traffic volume or roadway geometry is changed, or a new traffic control device is added, or a new traffic management scheme is implemented, they may cause some changes in the simulation results. These types of changes can be simulated, and the results can be compared to benchmark data, such as done in a before-and-after study, with and without implementing a change, the effect of changes in traffic parameters, or changes in roadway geometry.

8.1.1.1 *Example of External Variability*

An obvious example of external variability sources is changing traffic input values or system operation conditions. For example, effects of traffic volume increase, change in the number of lanes open to travelers, pedestrian volume at an intersection, or any driver, vehicle, or roadway characteristic.

It is possible that the internal variability is larger than the external variability. This was observed in an arterial simulation done by the author that showed the internal variability of the simulation results was larger than adding 10% more traffic to the arterial.

8.1.2 *Internal Variability Sources*

The internal variability is caused by the simulation model without changing any real-world traffic or roadway variables. The numerical values for input variables (such as traffic volume, roadway geometry, and driver characteristics) did not change, but the output may change due to the sequence in which the stochastic variables are generated (arranged). This is a little hard to understand, and I hope the following examples help you to comprehend it.

8.1.2.1 *Example 1 for Internal Variability*

Let's say you are studying the effects of average driver age on start-up lost time (the time at the beginning of green signal indication that is not used efficiently) at a signalized intersection. Say you are using the first four drivers for this purpose. In one simulation run, the first four drivers are labeled as Older, who is more cautious and reacts after making sure it is safe to accelerate slowly, so it takes 4 sec to roll over the stop bar. All drivers behind him/her must wait for this driver to move ahead. The second driver is a Middle-aged person who is a somewhat cautious driver, reacts with some pause to make sure that it is time to go, and then accelerates at a normal rate when it is possible. So, the time to go through the intersection for this driver is 3 sec. The third and fourth drivers are Younger drivers who react quicker than the Older and Middle-aged drivers and accelerate at a higher rate than the other two drivers. So, it takes 2 sec to go through the intersection. Now, the sequence of arrival of these drivers at the intersection is different in the two simulation runs:

- Run 1, the sequence of drivers is Older, Middle-aged, and the last two drivers are the Younger ones.
- Run 2, the sequence of drivers is Younger, Younger, Middle-aged, and the last driver is the Older driver.

The average headway in both cases is 11 (4 + 3 + 2 + 2 = 11 or 2 + 2 + 3 + 4 = 11) seconds. However, the time-in-queue experienced by those drivers in the early partition of the green time is different for Run 1 and Run 2. Thus, the delay the drivers experience is different even though the average headway is the same. For simplicity, let's assume time-in-queue is equal to delay (a more realistic approach would be assuming a portion of it is delay, but it does not change the nature of this discussion). So, in Run 1, the delay for car 1 is 4 sec, for car 2 it is 7 sec, for car 3 it is 9 sec, and for car 4 it is 11 sec. Total delay for those 4 cars is 31 sec (4 + 7 + 9 + 11 = 31). However, for Run 2, it is 24 sec (2 + 4 + 7 + 11 = 24). Note that there was no geometry or traffic volume change in these two runs. In both runs, we processed four vehicles, but the time needed to process them was 31 sec in one case and 24 sec in another case. What caused this is the arrangement of drivers at the queue at the intersection, which is done randomly. This effect can be observed in the output of two simulation runs where every input variable is the same except that we used two different random seeds to generate the sequence of arrival of those four cars.

8.1.2.2 *Example 2 for Internal Variability*

Another example of internal variability could come from the sequence of vehicles following each other in car following situations. Different simulation runs with different random number seeds may produce different results. Let's say you used different random number seeds and generated four vehicles in the sequences shown in Figure 8.1:

In both cases, we have four vehicles to move with a car following logic, but the car following behaviors of the group of four vehicles may not be the same.

8.1.2.3 *Example 3 for Internal Variability*

Let's assume we have nine vehicles of which three are tractor-trailers (T) and six are cars (C). We know that the headway between successive vehicles varies depending on who is following whom. A steep upgrade section of the road slows down the trucks more than cars, resulting in larger headways between trucks and cars compared to cars and cars. In real-world traffic, one cannot control how trucks will mix with cars. Thus, all possible combinations of arranging three trucks with six cars may happen. This implies that one must consider multiple sequences, and not just one, to reflect the real-world traffic conditions. It is theoretically possible but practically infeasible to find all possible arrangements for the trucks and cars on the platoon. In practice, you don't need to simulate all

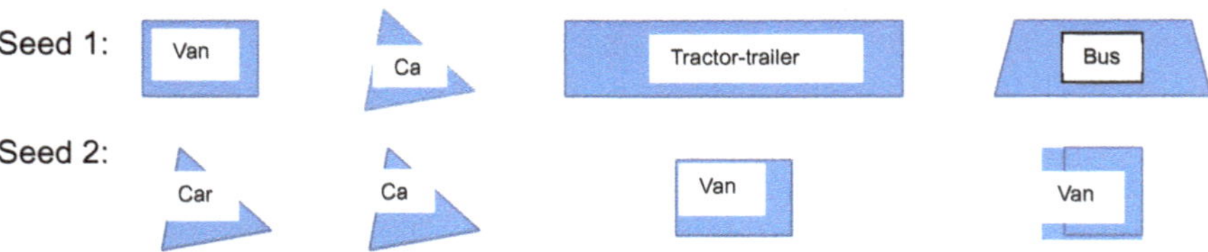

Figure 8.1. Effect of random number seed in sequence of vehicles generated.

Table 8.1. Effect of random number seed on sequence of vehicles and headways.

Seed Number (Run Number)	Sequence of Vehicles in Platoon		Average Headway (sec)
Seed 1	*C C C T T C C C T*	→	2.6
Seed 2	*C T C C T C T C C*	→	3.2
Seed 3	*T T C C T C C C C*	→	2.4
....	....		....
Seed *n*	*T T C T C C C C C*	→	2.3

possible combinations of the arrangements. We discuss later how many simulation runs should be enough to represent the entire combination.

Let's assume the sequence of vehicles are as shown in Table 8.1, and their corresponding average headways. With Seed 1, the lead vehicle is a truck followed by three cars, followed by two trucks, and at the tail end of this platoon, we have three cars. The average headway of these nine vehicles is 2.6 sec. The average headway for Seed 2 is 3.2 because trucks are intermixed with cars, and in Seed 3, cars are following each other in most cases, and the average headway is 2.4 sec.

There are nine vehicles in each platoon, regardless of the seed number; however, the behavior of a nine-vehicle platoon will be different depending on the seed number used.

There are two main approaches to deal with internal variability: (a) replication, (b) batch means. These topics and the number of runs are discussed in the following sections.

8.1.3 *How to Deal with Variability*

To deal with the internal variability in the output of simulation models, generally, two approaches are used: (a) batch means and (b) replications.

8.1.3.1 *Batch Means Approach*

In the batch means approach, you make one long simulation run and divide it into many smaller time intervals (batches). With this approach, there is only one transition period (warm-up time) at the beginning, then batches start. Data are collected for each batch. The data variability among the batches is used to assess the confidence level on simulation output. The number of batches (k) and the length of each batch (m) (also referred to as batch size) need to be decided (Banks *et al.*, 2001). Also, correlations between adjacent batches and time series relations among batches are issues that must be paid attention to. The length of each batch should be long enough that two adjacent batches can be considered as not-significantly correlated (uncorrelated or correlation is weak). The adjacent batches may be slightly correlated, but that correlation should not be significant (strongly correlated). Batches that are not adjacent should not show a time series trend. A time series trend would be when e.g. the average values of batches 5, 10, 15, and 20 are notably higher than the average values of other batches. If this happens, the length of the batch needs to be changed so that the cyclic pattern is observed. There is no widely accepted simple method to determine batch size m and consequently number of batches K, but Banks *et al.* (2001) gave some general guidelines on how to deal with batch size and autocorrelation.

The advantage of using the batch means concept is that there is only one transition time. This is very beneficial for the systems that require a long transition period. The disadvantage of the batch means approach is that the number of batches, the length of a batch, and the correlation among batches must be determined.

8.1.3.2 *Replication Approach*

In the replication approach, the length of a simulation run is much shorter compared to the length of a batch means run. Rather than one long run in a batch means approach, many independent shorter runs are made using different random number seeds. Thus, the autocorrelation that may exist in the batch means method is not an issue with the replication method because there are no batches. However, one must know the number of replications and the length of each run, as well as the length of the transition period. In each replication run, there is a transition period. For a system that has a short transition period, the replication approach is more suitable than the batch means approach. The mean, variance, and other indicators may be used to find the average performance of the replications. In traffic simulation, the replication method is used more often than the batch means, mainly because it is conceptually less complicated. In the past, many authors have made multiple runs (replication method) in their traffic simulation study; however, they did not address how

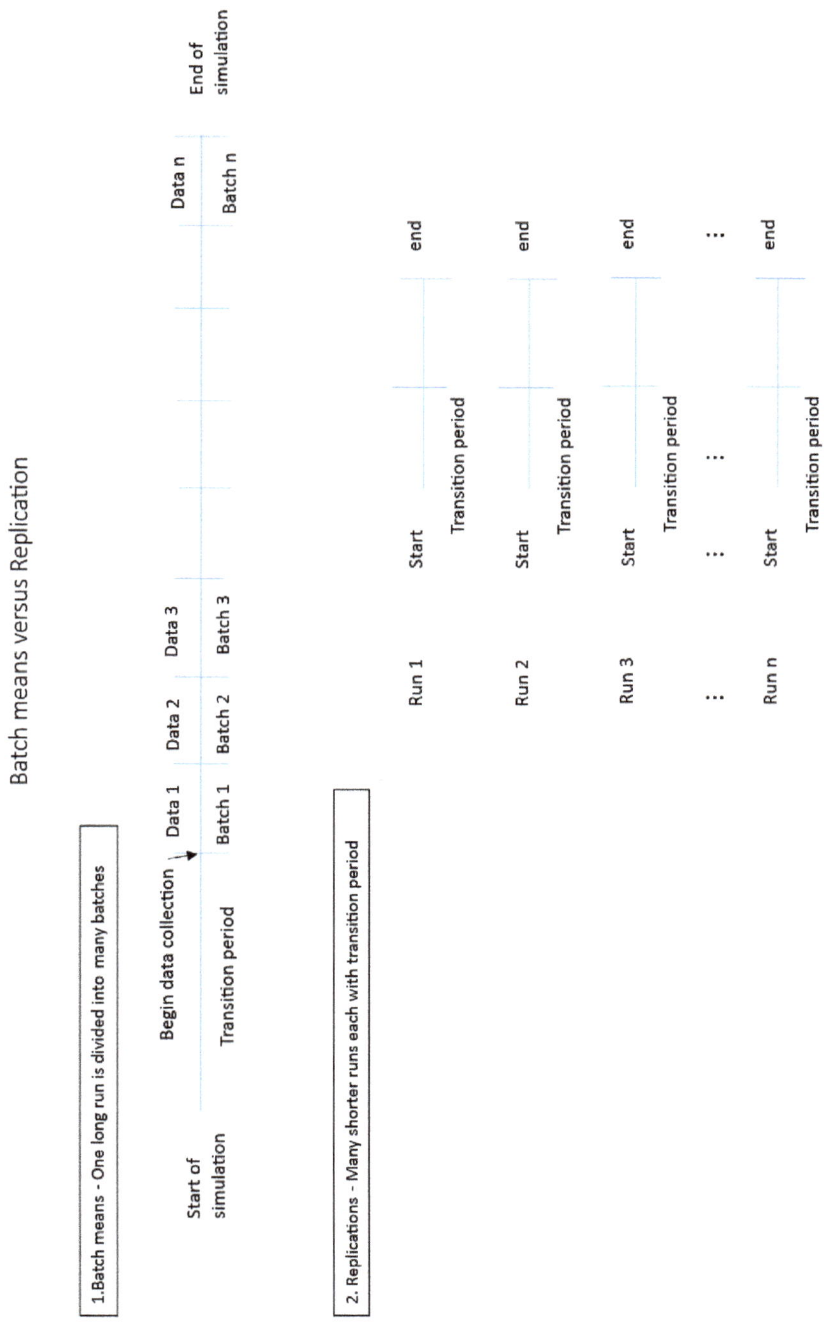

Figure 8.2. Batch means vs replication approach.

successful they were in dealing with the variability issue (Bruce & Hummer, 1991; Sadegh & Radwan, 1988; Radwan & Hatton, 1990; Kim & Messer, 1992; Torres *et al*., 1986).

Schematically, these two approaches are shown in Figure 8.2.

8.1.3.3 *Example Problem for Determining Number of Replications*

Suppose a simulation study resulted in an average delay of 52.0 ($\bar{x} = 52.0$) sec/veh and a standard deviation of 9.5 ($s = 9.5$) sec/veh. Desired margin of error = 8% of the mean at 95% confidence ($t \approx 2.201$):

$$e = 0.08 \times 52.0 = 4.16$$

$$n = (t \cdot s/e)^2 = (2.201 \times 9.5/4.16)^2 = (5.026)^2$$

$$n = 25.262 \rightarrow 26 \text{ replications required}$$

8.1.4 *Selection of Batch Means or Replication*

A few criteria may be used to decide which one to use. The first criterion should be the knowledge and familiarity of the user with statistical techniques. If the user has a limited statistical analysis background, one should consider the computer resources needed. If a single run of the simulation takes days to run it with the batch means approach and transient time is considerable portion of it, consider running replications. Also, the software may already have an approach implemented, then it may be easier to use whatever is implemented in the software. Some software allows the user to input the number of replications to run.

8.1.5 *Relationship between Internal and External Variabilities*

There is no direct relationship between internal and external variability that can be generalized to all simulation models. Internal variability shows how the simulation model is sensitive to the arrangement of (sequence of) random numbers generated and used in the simulation model to handle the effects of stochastic variables. While external variability is showing the effects of changes in the input parameters on the outcome. In a simulation study using VISSIM, the author compared the magnitude of the internal variability to the external variability. In a corridor with three signalized intersections, it was found that the internal variability caused about 7% change in delay, which was equivalent to a volume change of about 10%. The range of changes at different intersections and links was 6.99–13.6% changes in the delay, which was equivalent to a 6.1–18.9% increase in volume.

8.1.6 *Initial Transient State*

When a simulation starts, the system may be "empty" and it may take time to reach a "full" state. While the system is getting from an empty state to a steady state ("full"), data collected in this period may not represent the system properly. The data should be collected after the system reaches a steady state condition. Some refer to this time as start-up time, warm-up time, initial condition, truncation period, or period to reach an equilibrium condition. The data before the steady state conditions (initial transient state) should not be used to minimize the bias on the response variable. There is no definite rule on how the bias should be eliminated and how much of the simulation run should be truncated for this purpose. One way of finding out how much of the data should be truncated is by plotting the response variable vs time and locating the beginning of the region of steady state conditions. Discussion of various truncation policies is presented elsewhere (Benekohal, 1991), and a short summary is presented here.

The start-up policies to reduce the effects of the initial transient state were surveyed by Wilson and Pritsker (1978a). They reported in another study (Wilson & Pritsker, 1978b) that deleting data from the beginning of the simulation output to reduce initial transient effect (bias reduction) causes loss of information and an increase in the variance. Considering bias, variance reduction, and mean square errors, they suggested starting as close to the steady state mode as possible and keeping all data. Furthermore, they stated that "the judicious selection of an initial condition appears to be more effective than truncation in improving the performance of the sample mean as an estimator of the steady-state mean".

Some authors have advocated that for small and well-behaved models, the truncation should not be done, but for large models, this may not be the case (Pritsker & Pegden, 1979). Even after truncating the transient state, the simulation results will continue to fluctuate due to the stochastic variable used in the model.

The rule of thumb is that you might discard the observations as long as the response variable continues to increase or decrease. Kleijnen (1977) discussed that replication of simulation runs is the only alternative for gathering statistics about terminating systems. In a terminating system, the simulation run ends if a specific event occurs. To obtain independent replications, different random number seeds ought to be used.

8.1.7 *Variance Reduction Techniques*

After running replications of a simulation model, one can get the mean and variance of the responses. If the outcomes of the replications are considered to be independent, one can run a *t*-test and determine the confidence level for the mean of the observations. However, in simulation, the results do not always meet the independence requirements of the *t*-test or *f*-test. Kleijnen (1976) gave techniques to compare means and variances of two simulations where the outcomes from replications are pairwise *correlated. To compare the means of autocorrelated observations of two simulation experiments*, Fishman (1967) suggested that the difference of the sample means be treated as a normal variate for a sufficiently long sample record. Although this procedure is not as good as comparing autocorrelation structure of the two experiments, it suffices as an initial step for comparison (Kleijnen, 1976; Fishman, 1967).

There are two established variance reduction techniques when comparing two or more scenarios: (1) common random numbers (CRN), also known as correlated sampling, and (2) antithetic variates (AV). In CRN, the idea is to use the same sequence of random numbers in generating random characteristics (such as driver reaction time and desired speed) of one system, so they are similar to the characteristics in another system. This technique creates outcomes that are positively correlated. When the difference in performance of two scenarios is computed, this positive correlation should induce a negative correlation between the difference of the two outcomes, and that reduces the variance of the difference.

Mathematically, the relationship for the variance of $(X - Y)$ is

$$\mathrm{Var}(X - Y) = \mathrm{Var}(X) + \mathrm{Var}(Y) - 2\mathrm{Cov}\,(X, Y)$$

By introducing a positive correlation between X and Y, the covariance of X and Y (Cov (X,Y)) will be greater than zero. So, the variance of the difference $(X - Y)$ will be reduced by twice the amount of covariance values.

The antithetic variate technique is based on creating a negative correlation between pairs of observations of the same system to reduce the variability in the simulation outcome. Let's say we used random numbers U_1, U_2, U_3, …, U_n for the original run of the simulation. Let's say the outcome is Y_1. A corresponding "antithetic" simulation run is made using $(1 - U_1)$, $(1 - U_2)$, $(1 - U_3)$, …, $(1 - U_n)$. Let's call it Y_2. The average of the outcomes from run 1 and run 2 is $(Y1 + Y2)/2$. The variance of the average is

$$\mathrm{Var}((Y1 + Y2)/2)) = (1/4)[\mathrm{Var}(Y1) + \mathrm{Var}(Y2) + 2\mathrm{Cov}\ (Y1,\ Y2)]$$

With the negative correlation created, the $\mathrm{Cov}(Y1, Y2)$ will be less than zero. Thus, the variance of the estimator is reduced.

When two operating conditions are to be compared, one can introduce a negative correlation between replications of runs under one operating condition to reduce the variance for within runs and then introduce a positive correlation between runs under different operating conditions to reduce the variance for the difference between runs. This procedure is suggested as an efficient experiment design if there is no initial transient phase (Emcshoff & Sisson, 1970). This variance reduction method uses antithetic variate and common random numbers jointly. Kleijnen (1975) discussed possible undesirable effects of such a combination and indicated that because of cross-correlations the results may even be worse than using each method alone. He compared the three methods for different conditions and concluded that one cannot say which method is best for all systems.

Application of CRN for variance reduction in the NETSIM traffic simulation model was carried out by Rathi and Santiago (1990). Also, Rathi and Venigalla (1992) used the antithetic variate technique to reduce the variance of simulation outcomes.

8.1.7.1 *Example Problem for Common Random Numbers* (*CRN*)

Suppose you are conducting a simulation study to compare alternative X to alternative Y. Assume $\mathrm{Var}(X) = 100$ and $\mathrm{Var}(Y) = 64$. Covariance between X and Y plays a significant role. Three cases are examined:

Case 1: $\mathrm{Cov} = 0 \rightarrow \mathrm{Var}(X{-}Y) = 100 + 64 = 164$.
Case 2: $\mathrm{Cov} = +40 \rightarrow \mathrm{Var}(X{-}Y) = 100 + 64 - 2(40) = 84$.
Case 3: $\mathrm{Cov} = -20 \rightarrow \mathrm{Var}(X{-}Y) = 100 + 64 - 2(-20) = 204$.

CRN aims to create positive covariance so variance of the difference decreases, as is in Case 2.

8.1.7.2 *Example Problem for Antithetic Variate (AV)*

Suppose you are conducting a simulation study and utilizing the antithetic variates concept. Let's assume Var($Y1$) = Var ($Y2$) = 49. Assume that Cov($Y1$, $Y2$) = −21:

> Var(($Y1$ + $Y2$)/2) = (49 + 49 + 2(−21))/4 = 14
> Without AV (Cov = 0): Var(($Y1$ + $Y2$)/2) = 24.5
> When the variance is small, the number of runs would be smaller.

Antithetic Variates—Mild Negative Correlation Case

> Var($Y1$) = Var($Y2$) = 36, Cov($Y1$, $Y2$) = −3
> Var(($Y1$ + $Y2$)/2) = (36 + 36 + 2(−3))/4 = 16.5
> Without AV: Var(($Y1$ + $Y2$)/2) = 18

8.1.8 *Correlation among Batches*

When a long run is divided into many batches, the statistics for these batches may be correlated. A visual approach to see if the time series is stationary is to plot the output variables for the batches against time. Also, analytical techniques can be used to determine whether a time series is stationary. In time series data, the correlation coefficient between a variable in different batches is called autocorrelation. In a stationary time series data, the mean and variance are essentially constant over time. The values of the variable may come from two adjacent batches (lag 1) or may come from batches that are k time periods apart (called lag k). We assume a time series of Z_b, Z_{b+1}, …, Z_m is stationary. The autocorrelation coefficient at lag k, denoted by r_k, is computed from the following simplified equation (Bowerman & O'Connell, 1987; Chatfield, 1996):

$$r_k = \frac{\sum_{t=b}^{n-k}(z_t - \bar{z})(z_{t+k} - \bar{z})}{\sum_{t=b}^{n}(z_t - \bar{z})^2}$$

$$\bar{z} = \frac{\sum_{t=b}^{n}(z_t)}{(n-b+1)}$$

where Z_t is the value of the variable of interest in batch t, n is the number of batches, b is the batch number where the data collection starts from.

Here, r_k measures the linear relationship between time series observations separated by a lag of k time units. The value of r_k takes a value between −1 and +1, and when it is close to −1 or +1, it means the observations separated by k time units have a strong tendency to move together in a linear fashion (Bowerman & O'Connell, 1987).

An approximation for the standard error of r_k is

$$S_{rk} = \frac{\sqrt{1 + 2\sum_{j=1}^{k-1}(r_j)^2}}{\sqrt{(n-b+1)}}$$

Then, one can calculate the t statistic for autocorrelation at lag K, t_{rk}, by the following equation:

$$t_{rk} = \frac{r_k}{S_{rk}}$$

If $|t_{rk}| > t_{\alpha/2,\text{df}}$, you can reject the null hypothesis that $r_k = 0$ at a given significance level α and a known degree of freedom. Here, df is equal to $(n - b - 1)$ because two parameters are estimated (the mean of the time series and the autocorrelation coefficient). For large $(n - b + 1)$, the t_{rk} is often compared to a standard normal (Z) distribution instead of a t-distribution. For the standard normal distribution, the critical values at 5% α level are +1.96 and −1.96.

8.1.8.1 *Example Problem for Auto Correlation of Lag 1*

Suppose a batch means study produces the following average delay (sec/veh) from 10 batches:

$$z = [50, 55, 60, 65, 70, 75, 80, 85, 90, 95]$$

Obviously, this sequence shows an increasing trend. We compute lag-1 autocorrelation using the equation given above:

Compute the mean:

$$\bar{z} = 725/10 = 72.5$$

Compute deviations and squared deviations:

t	z_t	$z_t - 72.5$	$(z_t - 72.5)^2$		
1	50	−22.5	506.25	$(z_{t-1} - 72.5)$	**Product**
2	55	−17.5	306.25	−22.5	393.75
3	60	−12.5	156.25	−17.5	218.75
4	65	−7.5	56.25	−12.5	93.75
5	70	−2.5	6.25	−7.5	18.75
6	75	2.5	6.25	−2.5	−6.25
7	80	7.5	56.25	2.5	18.75
8	85	12.5	156.25	7.5	93.75
9	90	17.5	306.25	12.5	218.75
10	95	22.5	506.25	17.5	393.75

Compute Squared Deviations

$$\Sigma (z_t - \bar{z})^2 = 2062.5$$

Compute Adjacent Deviation Products

$$\Sigma (z_t - \bar{z})(z_{t-1} - \bar{z}) = 1443.75$$

Step 4: Compute Lag-1 Autocorrelation

$$r_1 = 1443.75/2062.5 = 0.700$$

Interpretation: $r_1 = 0.700$ indicates strong positive autocorrelation. Adjacent batches move together, indicating non-stationarity and violation of independence assumptions. Confidence intervals based on these batches would likely be misleading unless the batch size is increased or the transient period is reassessed.

8.1.8.2 *Example Problem for Batch Means vs Replication*

Application of batch means and replication approaches to the NETSIM traffic simulation model was discussed by Benekohal and Abu-Lebdeh (1994), and their example is briefly presented here.

If batch means is used, one must be careful in computing the variables for each batch. If the simulation model gives cumulative values, directly dividing the outcomes into smaller intervals (batches) is not appropriate because this may create a strong correlation among batches. Instead, the Proposed Interval Calculation (PIC) method suggested by Benekohal and Abu-Lebdeh (1994) should be used to find the values presenting the results for that interval alone (and not the cumulative values).

The nine-intersection street network Benekohal and Abu-Lebdeh used is shown in Figure 8.3. Hourly traffic volumes and spacing of the intersection are shown on the graph. The traffic signals operated with a 60 sec cycle, and the overall network was not congested. The measures of effectiveness (MOE) used are average delay, average speed, and the number of vehicle trips. The delay time is the difference between the actual time that a vehicle spends in the system and the ideal time based on free-flow speed. The average speed is defined as the ratio of the total distance traveled by all vehicles in the system to their total travel time. A vehicle completes a trip on a link when it passes over the stop line at the end of the link. MOEs were gathered at the network, intersection, and link levels. They made 24 replications and compared the average values from the replications to the average values obtained from a long run that was divided into 24 batches. The comparison outcome is shown in Table 8.2 (this table is created based on a larger table in Benekohal and Abu-Lebdeh's paper). The outcomes from the replication method are significantly different in seven cases and not significantly different in two cases.

Table 8.2. Comparison of MOEs from batch means method to the results from replication method.

	Delay (sec/veh.trip)			**Speed (mph)**			**Vehicle Trips (veh.trips)**		
Method	**Network**	**Link 9-6**	**Inters. 6**	**Network**	**Link 9-6**	**Inters. 6**	**Network**	**Link 9-6**	**Inters. 6**
Replication	73.2	29.66	37.33	12.9	11.5	9.9	1110	473	108
BM/PIC	75.0	32.09	42.82	12.7	11.1	9.1	1123	475	109
Significantly Different	Yes	Yes	Yes	Yes	Yes	Yes	Yes	No	No

Source: From Benekohal, R. F., and G. Abu-Lebdeh. Variability Analysis of Traffic Simulation Outputs: Practical Approach for TRAF-NETSIM. Transportation Research Record: Journal of the Transportation Research Board, No. 1457, 1994, Table 3, p. 203. Copyright, National Academy of Sciences. Reproduced with permission of the Transportation Research Board.

Note: Yes (No): means replication results are (not) significantly different than both means results.

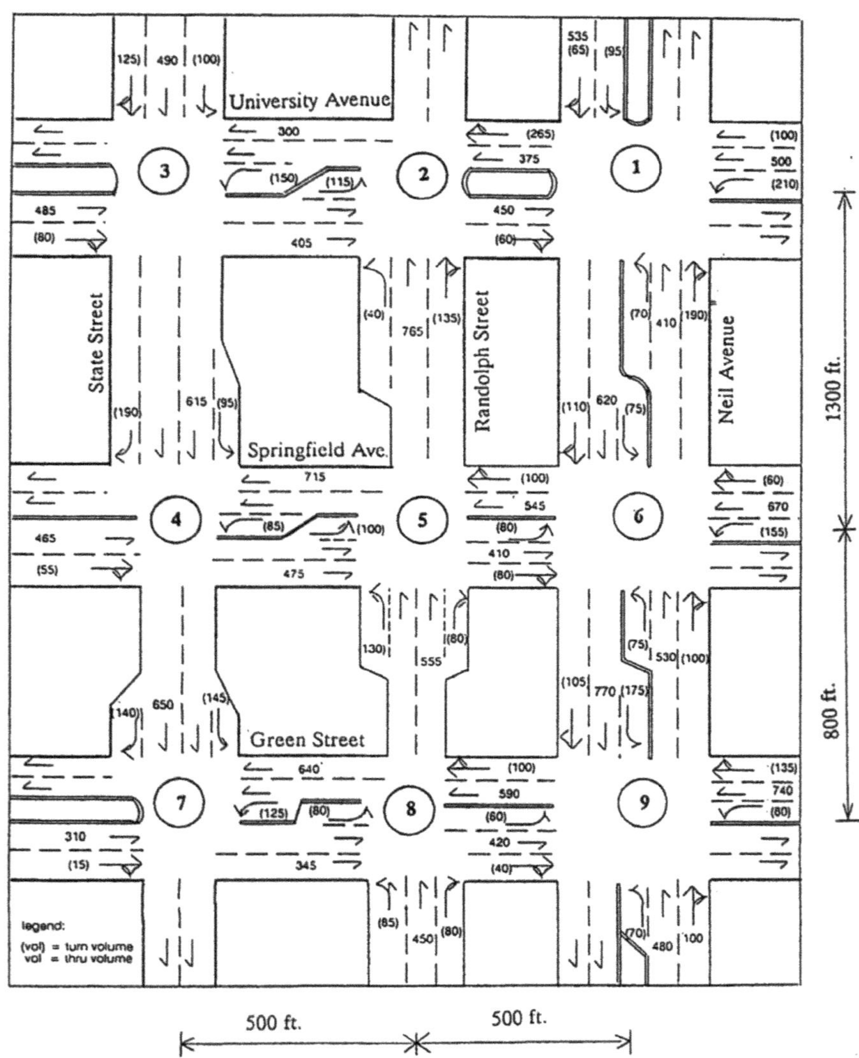

Figure 8.3. Network of intersections in city of Springfield, IL.

Source: From Benekohal, R. F., and G. Abu-Lebdeh. Variability Analysis of Traffic Simulation Outputs: Practical Approach for TRAF-NETSIM. Transportation Research Record: Journal of the Transportation Research Board, No. 1457, 1994, Figure 1, p. 201. Copyright, National Academy of Sciences. Reproduced with permission of the Transportation Research Board.

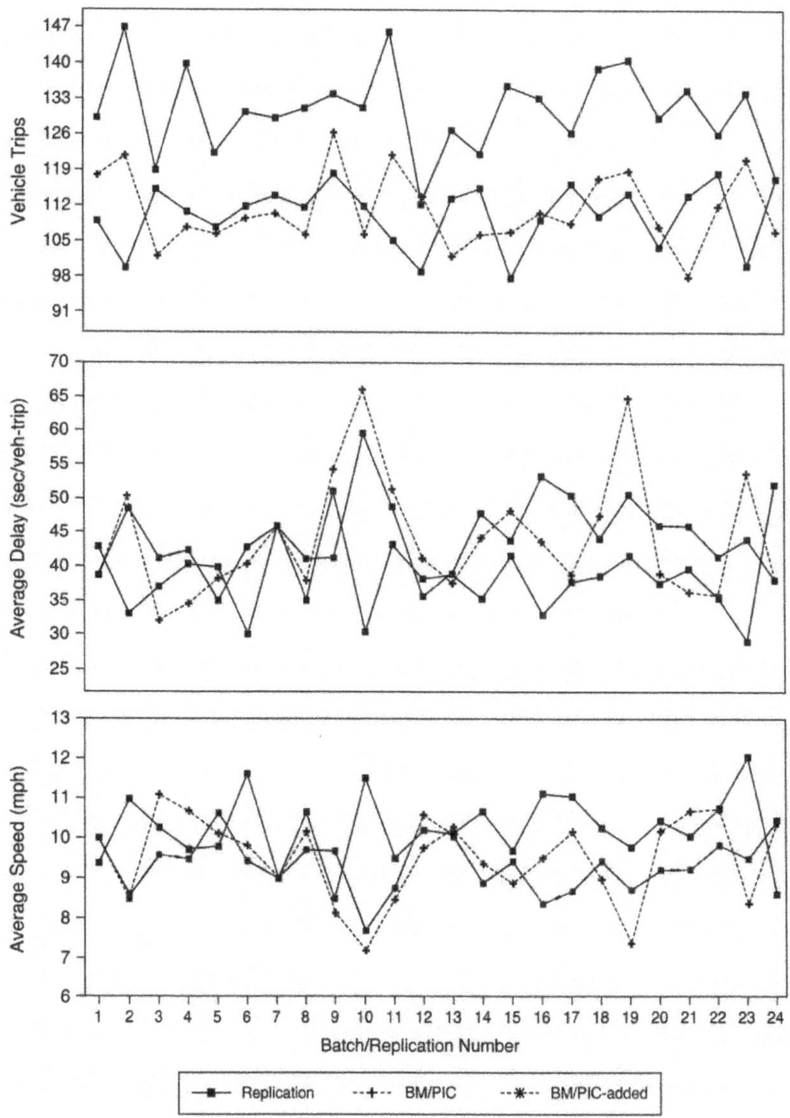

Figure 8.4. Comparison of speed, delay, and vehicle trips from replication vs batch means (BM/PIC) and Replications. BM/PIC-added is the outcome when 120 vehicles were added to Neil St northbound through traffic.

Source: From Benekohal, R. F., and G. Abu-Lebdeh. Variability Analysis of Traffic Simulation Outputs: Practical Approach for TRAF-NETSIM. Transportation Research Record: Journal of the Transportation Research Board, No. 1457, 1994, Figure 3, p. 205. Copyright, National Academy of Sciences. Reproduced with permission of the Transportation Research Board.

8.1.8.3 *Autocorrelation*

An example of autocorrelation for the average delay for a lag of 1 through 12 is shown in Table 8.3.

Examination of the r_k and s_{rk} values indicates that there was no strong correlation at any lag. Of particular interest is the correlation of adjacent batches (lag 1 values), which have t_{rk} values less than 1. The highest t_{rk} for the network was 1.2022, for intersection 6 was 1.3989, and for link 9-6 was 1.5814. These are all below the critical values for a *t*-test and indicate there was not a significant autocorrelation. The t_{rk} values greater than 1.6 are statistically large for lags of 1, 2, and perhaps 3 (Bowerman & O'Connell, 1987).

8.1.8.4 *Correlation Coefficient*

A simpler procedure to examine the correlation between adjacent batches (lag 1) is to compute the correlation coefficient (*r*). The correlation coefficient is one measure of the strength of the relationship between two variables (Ott, 1988). In the context of batch means, the correlation

Table 8.3. Autocorrelation for average delay for lags from 1 to 12.

Network					
Lag	r_k	S_{rk}	**Lag**	r_k	S_{rk}
1	.00939	.20412	2	–.08439	.20414
3	.12182	.20559	4	.06076	.20858
5	–.15485	.20931	6	–.14545	.21403
7	–.00879	.21811	8	–.01030	.21813
9	–.26227	.21815	10	–.05152	.23091
11	.17652	.23139	12	–.14364	.23694
Intersection 6					
Lag	r_k	S_{rk}	**Lag**	r_k	S_{rk}
1	–.14666	.20412	2	–.29162	.20847
3	.25560	.22482	4	–.05741	.23662
5	–.03321	.23720	6	–.07393	.23740
7	–.21173	.23835	8	–.00162	.24607
9	.08054	.24607	10	–.05383	.24716
11	.02738	.24765	12	–.07025	.24778

Table 8.3. (*Continued*)

Link 9-6

Lag	r_k	S_{rk}	**Lag**	r_k	S_{rk}
1	.15073	.20412	2	−.20663	.20871
3	−.18384	.21707	4	.14179	.22346
5	−.05707	.22718	6	−.36020	.22777
7	−.18585	.25039	8	.19916	.25607
9	.30415	.26244	10	−.01755	.27674
11	−.24191	.27679	12	−.05680	.28546

Source: From Benekohal, R. F., and G. Abu-Lebdeh. Variability Analysis of Traffic Simulation Outputs: Practical Approach for TRAF-NETSIM. Transportation Research Record: Journal of the Transportation Research Board, No. 1457, 1994, Table 2, p. 202. Copyright, National Academy of Sciences. Reproduced with permission of the Transportation Research Board.

coefficient would show how the output of batch 1 is related to the same output from batch 2, the outcome of batch 2 to the outcome of batch 3, that of batch 3 to that of batch 4, and so forth. Recall that the correlation coefficient for a variable in the adjacent batches is the same as the autocorrelation coefficient at lag 1. A simple formula for the autocorrelation coefficient at lag 1, r_1, which is the same as the correlation coefficient (r), can be written as

$$r = r_1 = \frac{\sum_{t=2}^{T}(z_t - \bar{z})(z_{t-1} - \bar{z})}{\sum_{t=1}^{T}(z_t - \bar{z})^2}$$

where T is the number of batches and other variables are defined before.

This procedure is much simpler than the time series analysis but is limited to the correlations of adjacent batches. The correlation coefficient and the Lag 1 autocorrelation coefficient would provide the same results. Thus, one may compute the correlation coefficient (r) if the dependency between adjacent batches is considered. The r values will be between -1 and $+1$. An $r > 0$ indicates a positive linear relationship between adjacent batches, an $r < 0$ indicates a negative linear relationship between adjacent batches, and $r = 0$ indicates no correlation between adjacent batches (Ott, 1988). As you can see, finding r and interpreting it is much easier than autocorrelation analysis.

8.1.9 *Which Method Should Be Used*

The question is which method should be used: batch means or replication. Factors to be considered in selecting a method would include human resources, computer time (both initialization and simulation), size of the system, and previous experience with each method. For large networks in which the initialization period is likely to be longer, batch means seems more logical than the replication method.

The replication method does not have the autocorrelation issue that the batch means method may have. However, the number of replications and the length of each run need to be determined.

8.1.10 *Number of Replications*

The approach Benekohal & Abu-Lebdeh (1994) suggested is to start with X *runs* (say 10). The length of each replication run would depend mainly on the size of the network and traffic conditions. They proposed that for small networks, the simulation time should be at least 10 signal cycle lengths. They also suggested that the data collection period beyond the initialization should be longer than the initialization period. Then, find the mean and variance for those X runs. Select a confidence level (usually 90% or 95%). Then, follow these steps to find the number of runs (n) needed:

1. From the data for X runs, compute the mean and variance, s^2.
2. Calculate n from the following equation:

$$n = (t * s)^2/e^2$$

where s is the standard deviation of the population, estimated by finding the standard deviation of the sample, e is the tolerable error, desired margin of error, or the maximum acceptable difference between the sample mean and the true population mean, and t is the *critical t-value from a* t-distribution with df of $n - 1$.

3. Compare n to X.
   a. If $n > X$, make a few additional runs and repeat the above procedure.
   b. If $n < X$, stop; your sample size is enough.

The tolerable error depends on the desired accuracy. To start the calculations, you may assume e is 10% of the mean. When there are extreme values in the observed data, first discard them and then find the range of data. Using 5–15% of the range was suggested by Benekohal and Abu-Lebdeh (1994).

For the batch means method, the length of the simulation run, *T*, needs to be found. Benekohal and Abu-Lebdeh (1994) suggested that *T* should be divided into *X* intervals, each being *T*/*X*. They suggested *T*/*X* be equal to an integer value of the cycle length for signalized networks, say, 10 average cycle lengths (if cycle length is 60 sec, starting *T*/*X* would be 10 min) and greater than initialization time. Run the simulation and compute MOEs for the *X* batches. Examine autocorrelation among batches, and if it is not significant, use this batch size. To determine the number of batches, use the procedure described in the replication section. If autocorrelation is a problem, increase the batch size. Fishman suggested doubling the batch size to expeditiously arrive at an appropriate batch length (Fishman, 1967). However, one may consider increasing batch size at a slower rate, perhaps by 50% at a time. Knowing the batch size and number of batches, one can easily find the length of the simulation run. If the autocorrelation problem persists and the batch length becomes unreasonably long, consider using the replication method.

8.2 Random Number Generator (RNG)

The ability to generate random numbers is considered the building block of a simulation study (Ross, 2002). There are many ways of generating random numbers. Simple examples for it would be tossing a dice to get a number between 1 and 6, asking someone to pick a number between 1 and 50 randomly, or selecting a number from a random number table (like the RAND corporation million-number table). Simulation packages need to generate independent identically distributed (IID) random numbers uniformly distributed between 0 and 1, $U(0,1)$, and are used to define the stochastic characteristics of the system and its users. For example, we may categorize drivers in terms of their reaction times into 10 levels. Then, a random number between one and 10 is assigned to one of the 10 levels of reaction times for this driver.

8.2.1 *Liner Congruential Generator*

As computers became more available, algorithms were developed to generate random numbers in arithmetic or numerical ways (Law & Kelton, 2000). The most common type of RNG is liner congruential generator

(LCG) introduced by Lehmer in 1951. Some people call these pseudo-random numbers because we can determine the sequence in which they are generated. The LCG is an arithmetic generator based on the following recursion algorithm:

$$S_i = (aS_{i-1} + \mathrm{b})\ (\mathrm{mod}\ C)$$
$$U_i = S_i/C$$

where S_0 is the starting value or the seed, a is the multiplier, b is the increment, and C is the modulus.

In this method, you divide $(aS_{i-1} + b)$ by C and let S_i be the remainder of the division. Then, compute S_i/C. This yields a random number between zero and 1. These numbers represent a uniform distribution between 0 and 1, $U\,(0,1)$.

8.2.1.1 *Example of an RNG*

Assume the starting seed, s_0, is 10, $a = 9$, $b = 7$, $C = 4$

$$S_1 = (9 \times (10) + 7) \mathrm{mod}\ 4 = 1$$
$$S_2 = (9 \times (1) + 7) \mathrm{mod}\ 4 = 0$$
$$S_3 = (9 \times (0) + 7) \mathrm{mod}\ 4 = 3$$
$$S_4 = (9 \times (3) + 7) \mathrm{mod}\ 4 = 2$$
$$S_5 = (9 \times (2) + 7) \mathrm{mod}\ 4 = 1$$
$$S_6 = (9 \times (1) + 7) \mathrm{mod}\ 4 = 0$$
$$S_7 = 3$$
$$S_8 = 2$$

For this example, S_i/C are 0/4, ¾, 2/4, and ¼.

This RNG is showing a "looping" behavior. After the first 4 numbers, it generates the same numbers again. This RNG can generate only four random numbers.

8.2.2 *LGC Constraints and Period*

A good RNG should not loop, or looping should occur after a very huge number of generations. To avoid quick looping, certain constraints are assigned to a, b, and c values. a, b, C, and S_0 must be non-negative integers, and $C > 0$, $C > a$, $C > b$, and $C > S_0$ (Law & Kelton, 2000). Even with these constraints, one may not have an LCG with good performance. For example, assume $C = 4$, $a = 3$, $b = 1$, $S_0 = 2$. So, all the constraints are satisfied, but the numbers generated are not that desirable:

- $S_1 = (3 \times (2) + 1)(\text{mod } 4) = 3$
- $S_2 = (3 \times (3) + 1)(\text{mod } 4) = 2$
- $S_3 = (3 \times (2) + 1)(\text{mod } 4) = 3$
- $S_4 = (3 \times (3) + 1)(\text{mod } 4) = 2$

This LCG looped after generating two numbers.

The length of the (looping) cycle is called the period (or cycle length). An LCG has a full period if it yields C values before looping. To have a long period, C must be a large number, say 10^9 (Law & Kelton, 2000).

To increase the efficiency of random number generation in computers, C should be a large prime number that can fit into the computer word size. Ross (2002) suggested using $2^{31}-1$ for a 32-bit word machine where the first bit is a sign bit. If LCG has a partial period, then the cycle length depends on the seed (S_0) used. For an LCG to have a full period, the following three conditions must be satisfied (Law & Kelton, 2000; Hull & Dobell, 1962).

Full period theorem proposed by Hull & Dobell (1962) is as follows:

(1) The increment b is relatively prime to modulus C.
(2) If q is a prime number that divides C, then q divides $a - 1$.
(3) If 4 divides C, then 4 divides $a - 1$.

8.2.3 *Types of LCG*

An LCG with $b > 0$ is called a mixed LCG. An LCG with $b = 0$ is called a multiplicative LCG. A multiplicative LCG will not have a full period because it does not satisfy condition 1 of the full period theorem (Law & Kelton, 2000). An LCG that combines two or more separate LCGs to get a final random number is called a composite LCG. For more discussions on these LCGs, one may refer to Law and Kelton (2000).

8.2.4 *RNG Used by Excel*

In a Microsoft Excel spreadsheet, a random number between 0 and 1 is generated using the RAND() function. Since Excel 2010, the RNG algorithm has changed from the middle-square method to the Mersenne Twister (MT) method. The middle-square method was developed by von Neumann and Metropolis in 1949 (Wikipedia). The algorithm takes an n digit number (n should be even) and squares it. Then, it treats the middle n digits of the new number as a random number. To make the number less than one would put a decimal point in front of it. Later, it was discovered that it produces patterns and could even get stuck repeating the same numbers (Law & Kelton, 2000). The MT algorithm generates uniform pseudorandom numbers and has a tremendously large period of $2^{19937}-1$, while consuming a working area of only 624 words (Matsumoto & Nishimura, 1998). This period is so large that for writing it down, one needs 6001 digits (it is approximately equal to 1.34×10^{6000}). The sequence is 623-dimensional equidistribution with up to 32-bit accuracy).

8.2.5 *Generating Random Variates*

What is the difference between a random variable, a random number, and a random variate? These three terms are sometimes used incorrectly. They are not interchangeable, though they are related. A random variable is a mathematical function that assigns a numerical value to each outcome of a random experiment. It is not a single number but a rule that yields an outcome. For example, if playing backgammon, you roll two dice, and each dice shows a face that may have 1 through 6 dots. If we define X as the number of faces with 6 dots. Its values would be 0 (indicating no face with 6 dots), 1 indicating one of the dice shows 6 dots, and 2 indicating both dice show 6 dots. So, X is a random variable that can take values of 0, 1, and 2. X has a distribution with a mean and a variance. A random variate is the outcome of the random variable in a trial. In this example, a random variate may take a value of 0 in one trial. In another trial, it may have a value of 1, and so on. A random number is a general term and has a numeric value produced by a stochastic process or generated by an algorithm, such as RNG algorithms. The difference between a random number and a random variate is that a random number is an arbitrary value produced by an RNG, but a random variate is a value drawn from a probability distribution.

There are several algorithms to generate random variates from a given distribution. The exactness, efficiency, and complexity of the algorithms must be considered in selecting them (Law & Kelton, 2000). One of the popular algorithms is inverse transformation, where a given random number U (a random number from a uniform distribution between 0 and 1) results in a random variate from a distribution. Other techniques such as compositions, convolution, acceptance-rejection for continuous distributions, as well as the algorithms for discrete distributions are discussed in simulation books (such as Law & Kelton, 2000; Banks *et al.*, 2001).

Exercises

Part 1. Questions

1. Define external variability and internal variability in simulation and give two examples of each from traffic simulation.
2. How can we gauge the magnitude of internal variability?
3. This chapter notes that internal variability can be larger than external variability. Explain how this can happen and what it implies for study design.
4. If a system does not have any stochastic variable, is internal variability still an issue?
5. What are the two main approaches to dealing with the output variability in simulation models?
6. When is the batch means method more appropriate than the replication method?
7. What should we not use the results coming from the initial transient state in system performance analysis?
8. What is the difference between the tolerable error "e" in the sample size equation and the error (inaccuracy) in measuring field data?
9. Why are we concerned about the period of a random number generator?
10. In an Excel spreadsheet, how many random numbers can be generated before the numbers are repeated?
11. Is a random variable the same as a random variate?
12. This chapter states that seemingly constant outputs across runs should not automatically be interpreted as "steady state". Explain why.

13. Define the initial transient state and explain why truncation is used.
14. What is a terminating system? And why are replications often required for terminating systems?
15. Compare the replication and batch means approaches in terms of
    a. Initialization/warm-up handling
    b. Independence/autocorrelation concerns
    c. Practical ease of use in traffic simulation
16. What does it mean if the lag-1 autocorrelation in batch means is large and positive?
17. Explain the goal of variance reduction techniques and why they matter when comparing scenarios.

Part 2. Problems

Problem 1. Internal variability example (driver sequence).
This chapter's Example 1 has two runs with four vehicles and processing times:

- Run 1 headways: 4, 3, 2, 2 seconds.
- Run 2 headways: 2, 2, 3, 4 seconds.

Assume these four vehicles were stopped at and assume there was no delay for them before the first vehicle started moving. Assume delay = time-in-queue in this example.

a. Compute delay for each vehicle and total delay for Run 1.
b. Compute delay for each vehicle and total delay for Run 2.
c. Compute average delay per vehicle in each run.
d. Explain why average headway is identical but delay differs.

Problem 2. Internal variability from truck/car ordering.
A platoon contains 9 vehicles (3 trucks T, 6 cars C). Suppose the average headways (sec) depend on follower–leader type:

- C following C: 2.0
- C following T: 2.5
- T following C: 3.0
- T following T: 3.5

Two sequences occur:

- Sequence A: $C\ C\ C\ T\ T\ C\ C\ C\ T$
- Sequence B: $C\ T\ C\ C\ T\ C\ T\ C\ C$

Assume headway is determined by each follower–leader pair; there are 8 headways in a 9-vehicle platoon.

a. Compute the average headway for each sequence.
b. Explain how random seed affects the sequence and, therefore, the output.

Problem 3. When internal variability "masks" external variability.
A corridor simulation shows that changing volume by +10% changes average delay by +6%. However, different random seeds (same inputs) cause delay to vary by ±8%.

a. Explain what this implies about detecting the effect of the +10% volume change.
b. Propose a strategy to still make a reliable conclusion (method + what to increase/adjust).

Problem 4. Batch means setup.
You are running a 1-hr simulation of an arterial. Warm-up is estimated as 10 min.

a. Propose a batch means plan with k batches of equal size after warm-up.
b. Choose a batch size m (minutes) and compute k.
c. Explain what conditions must be checked regarding correlation and time series trends.

Problem 5. Lag-1 autocorrelation calculation (simplified).
A batch means study yields an average delay for 8 batches:

$$z = [72, 75, 74, 76, 73, 74, 75, 73]$$

a. Compute $\bar{z}$.
b. Compute lag-1 autocorrelation $r1$ using the equation given in this book.
c. Interpret whether adjacent batches appear strongly correlated.

Problem 6. Determining number of replications (pilot study).
You run 10 replications (after warm-up removal) and obtain $\bar{x} = 38.0$ sec/veh and $s = 6.0$ sec/veh.

You want a margin of error $e = 10\% \cdot \bar{x}$ at 95% confidence. Use $t \approx 2.262$ (df ≈ 9) for an initial estimate. Are 10 runs enough?

Problem 7. CRN effect on variance of differences.
Two scenarios produce outputs X and Y for each replication. Sample estimates: Var(X) = 25, Var(Y) = 16.

Compute Var($X - Y$) for each of the following covariance values:

a. Cov(X, Y) = 0
b. Cov(X, Y) = +10
c. Cov(X, Y) = −10

Explain why CRN aims to make Cov(X, Y) positive.

Problem 8. Antithetic variates.
Two antithetic runs produce outputs $Y1$ and $Y2$. Assume Var($Y1$) = Var($Y2$) = 36 and Cov($Y1$, $Y2$) = −12.

Compute Var[($Y1$ + $Y2$)/2)] = [Var($Y1$) + Var($Y2$) + 2Cov ($Y1$, $Y2$)]/4.
Compare it to the variance when Cov ($Y1$, $Y2$) = 0.

Problem 9. LCG computation.
Given the linear congruential generator (LCG): $Si = (aSi - 1 + b) \bmod C$, $Ui = Si/C$.

Let $S0 = 10$, $a = 9$, $b = 7$, $C = 4$.

a. Compute $S1$ through $S8$.
b. Compute $U1$ through $U8$.
c. Identify the period and explain what "looping" means.

Problem 10. Full period theorem check (conceptual).
State the three Hull–Dobell conditions for the full period of a mixed LCG and explain, in words, what each condition is trying to prevent.

Problem 11. Bad LCG example.
Given $C = 4$, $a = 3$, $b = 1$, $S_0 = 2$:

a. Compute $S1$ through $S6$.
b. Find the period.

c. Explain why satisfying simple constraints (e.g. $C > a$, $C > b$) is not sufficient for quality.

Problem 12. Random variable vs random variate.
Using the backgammon example (rolling two dice, each dice has 6 sides, each side shows a number of dots that can be 1 through 6), let X be the number of dice showing six dots.

a. Define the random variable X and its possible values.
b. Give two examples of random variates of X.
c. Give an example of a random number U in (0,1) and explain how it differs from a random variate.

Problem 13. Inverse transform example.
Arrivals follow an exponential distribution with rate λ. Inverse transform: $T = -(1/\lambda)\ \ln(1 - U)$.

a. If $\lambda = 0.5$ veh/sec and $U = 0.20$, compute T.
b. If $U = 0.90$, compute T.
c. Explain why a larger U tends to produce a larger T in this formula.

References

Banks, J., Carson, J. S., Nelson, B. L., & Nicol, D. M. (2001). *Discrete-Event System Simulation* (3rd ed.). Prentice Hall, New Jersey.

Benekohal, R. F. (1991). Procedure for Validation of Microscopic Traffic Flow Simulation Models, TRR 1320, TRB, pp. 190–202, National Research Council, Washington, D.C.

Benekohal, R. F. & Abu-Lebdeh, G. (1994). Variability analysis of traffic simulation outputs: Practical approach for TRAF-NETSIM. *Transportation Research Record, 1457*, 198–207. TRB, NRC.

Bowerman, B. & O'Connell, L. (1987). *Time Series Forecasting: Unified Concepts and Computer Implementation.* Duxbury Press, Belmont, CA.

Bruce, E. L. & Hummer, J. E. (1991). Delay alleviated by left-tum bypass lanes. *Transportation Research Record, 1299*, 1–8. TRB, NRC.

Chatfield, C. (1996). *The Analysis of Time Series an Introduction* (5th ed.). Chapman & Hall, London, UK.

Emcshoff, J. R. & Sisson, R. L. (1970). *Design and Use of Computer Simulation Models*. McMillan Co., New York.

Fishman, G. S. (1967). Problems in the statistical analysis of simulation experiments: The comparison of means and the length of sample records. *Communication of the ACM, 10*(2), 94–99.

Fishman, G. S. (1978). *Principles of Discrete Event Simulation.* Wiley & Sons Inc., New York.

Hull, T. E. & Dobell, A. R. (1962). Random number generators. *SIAM Review, 4*, 230–254.

Kim, Y. & Messer, C. J. (1992). Traffic Signal Timing Models for Oversaturated Signalized Interchanges. Research Report 1148-2, Texas Transportation Institute, Texas A&M University, College Station.

Kleijnen, J. P. C. (1975). Antithetic variates, common random numbers and optimal computer time allocation in simulation. *Management Science, 21*(10), 1176–1185.

Kleijnen, J. P. C. (1976). Comparing means and variances of two simulations. *SCI Simulation, 26*(3), 87–88.

Kleijnen, J. P. C. (1977). Design and analysis of simulations: Practical statistical techniques. *SCI Simulation, 28*(3), 81–90.

Law, A. M. & Kelton, W. D. (2000). *Simulation Modeling and Analysis* (3rd ed.). McGraw-Hill Inc., New York.

Lehmer, D. H. (1951). Mathematical methods in large-scale computing units. *Annals of Computer Laboratory of Harvard University, 26*, 141–146.

Matsumoto, M. & Nishimura, T. (1998). Mersenne Twister: A 623-dimensionally equidistributed uniform pseudo-random number generator. *ACM Transactions on Modeling and Computer Simulation, 8*(Issue 1: Special Issue on Uniform Random Number Generation), 3–30.

Ott, L. (1988). *An Introduction to Statistical Methods and Data Analysis* (3rd ed.). PWS-Kent Publishing Company, Boston, Mass.

Pooch, U. W. & Wall, J. A. (1992). *Discrete Event Simulation: A Practical Approach.* CRC Press, Boca Raton, FL.

Pritsker, A. A. B. & Pegden, C. D. (1979). *Introduction to Simulation and SLAM.* John Wiley & Sons, New York.

Radwan, A. E. and Hatton, R. L. (1990). Evaluation tool of urban interchange design and operation. *Transportation Research Record, 1280*, 148–155. TRB, NRC.

Rathi, A. K. & Santiago, A. J. (1990). Identical traffic stream in the TRAFNET-SIM simulation program. *Traffic Engineering and Control, 31*(6), 351–355.

Rathi, A. K. & Venigalla, M. M. (1992). Variance reduction applied to urban network traffic simulation. *Transportation Research Record, 1365,* 133–143. TRB, National Research Council, Washington, DC.

Ross, S. M. (2002). *Simulation* (3rd ed.). Academic Press, San Diego.

Sadegh, A. & Radwan, A. E. (1988). Comparative assessment of 1985 HCM delay model. *Journal of Transportation Engineering, 114*(2), 194–208.

Sokolowski, J. A. & Banks, C. M. (ed.) (2010). *Modeling and Simulation Fundamentals: Theoretical Underpinnings and Practical Domains*. Wiley & Sons, New Jersey.

Torres, J. F., Halati, A., & Danesh, M. (1986). Impact of Lane Obstruction, Vol. 2. Research Report. FHWA Contract DTFH61-84-C-00064. JFT Associates; FHWA, U.S. DOT, February 1986.

Wikipedia. https://en.wikipedia.org/wiki/Middle-square_method.

Wilson, J. R. & Pritsker, A. B. (1978a). A survey of research on the simulation start up problem. *Simulation, 31*(2), 55–58.

Wilson, J. R. & Pritsker, A. B. (1978b). Evaluation of start up policies in simulation experiments. *Simulation, 31*(3), 79–89.

Chapter 9

Comparison Criteria and Confidence Interval

There are many ways of comparing simulation results to reliable field data or benchmark data to gain confidence that the simulation outcomes are valid. In the Validation chapter, how to conduct these comparisons was discussed. This chapter discusses the criteria (performance measures) to be used for such comparisons and the confidence level that can be attained on such comparisons, as well as the sample size needed. To build a confidence interval, we need to know the unbiased estimator of variance. First, a brief discussion on the mean and variance of sample data and their relations to the mean and variance of the population is presented.

9.1 Sample Mean and Variance

Simulation results may vary from one run to another. For example, assume simulation runs 1, 2, 3, …, n resulted in values of $x_1, x_2, x_3, \ldots, x_n$. If it is assumed that x_i (i.e. $x_1, x_2, x_3, \ldots, x_n$) are independent and identically distributed (IID) random variable (with defined mean and positive variance), then central limit theorem can be used to compute a confidence interval on the true mean of the population (here, population means the variable we are studying, i.e. x). Let the average value of the sampled x_i be $\bar{x}$ and the variance of them be s^2:

$$\bar{x} = \frac{\sum_1^n (X_i)}{n}$$

$$s^2 = \frac{\sum_1^n (X_i - \bar{x})^2}{n}$$

where $\bar{x}$ is the sample mean and s^2 is the sample variance.

$\bar{x}$ and s^2 could be the statistics for a variable (e.g. speed, travel time, queue, and delay) obtained from "*n*" runs of the simulation model. As the number of simulation runs (the sample size "*n*") increases, the confidence in these values, in general, will increase.

9.2 Population Mean and Variance

The true mean (μ) and variance of the population (σ^2) of the simulation runs are unknown to us. We estimate μ and σ^2 using the sample data. Expected value of *x*, shown as $E(x)$, is the average value of *x* and is called the population mean and normally shown as μ:

$$\mu = E(x)$$

The variance of the population of *x* values, shown as σ^2, is the average value of $(x - \mu)^2$ (Hogg & Tanis, 1977):

$$\sigma^2 = E[(x - \mu)^2]$$

Expand the equation:

$$\sigma^2 = E(x^2 - 2\,\mu x + \mu^2)$$

Using the distributive property of *E*, it can be written as

$$\sigma^2 = E(x^2) - 2\,\mu E(x) + \mu^2$$

Simplify it:

$$\sigma^2 = E(x^2) - 2\,\mu(\mu) + \mu^2 = E(x^2) - \mu^2$$

It can be written as

$$\sigma^2 = E(x^2) - [E(x)]^2$$

where μ is the population mean and σ^2 is the population variance.

The true values of μ and σ^2 are unknown to the simulation user (because we don't make enough runs to represent every member of the population), but they can be estimated from the mean and variance of the sample simulation runs. Often, we use the mean and variance of the sample to estimate the mean and variance of the population, as shown in the following:

$$\overline{x} \approx \mu$$

$$\sigma^2 \approx s^2$$

9.3 Unbiased Estimator of Variance

To compute the variance of "n" numbers in sample data, the following equation is used. The sum of differences-squared is divided by a divisor "n":

$$s^2 = \frac{\sum_1^n (X_i - \overline{x})^2}{n}$$

However, some authors divide it by (n – 1), as shown in the following (Hogg & Tanis, 1977; Snedecor & Cochran, 1989):

$$s^2 = \frac{\sum_1^n (x_i - \overline{x})^2}{n-1}$$

Should the divisor be "n" or "n – 1"? Both equations are correct, but they have different implications. Following the strict definition of variance of a sample dataset, the divisor should be n (it should be divided by n). However, when sample data are used to find an unbiased estimate of the variance of the population, the divisor should be "n – 1" (it should be divided by (n – 1)). A mathematical justification for doing that, given by Hogg *et al.* (2021), is as follows. We can start with this equation:

$$E\left(S^2\right) = \frac{1}{n}\left[\sum_1^n E\left(x_i^2\right) - nE(\overline{x}^2)\right]$$

Substitution yields

$$E\left(S^2\right) = \frac{1}{n}\left[\sum_{1}^{n}(\mu^2+\sigma^2) - n\left(\mu^2+\frac{\sigma^2}{n}\right)\right]$$

Simplification yields

$$= \frac{1}{n}\left[n\mu^2 + n\sigma^2 - n\mu^2 - \sigma^2\right]$$

$$= \frac{1}{n}\left[(n-1)\sigma^2\right]$$

So, we have

$$E\,(S^2) = (n-1)\,\sigma^2/n$$

It can be written as

$$E\left(\frac{nS^2}{n-1}\right) = \sigma^2$$

That is,

$$\left(\frac{nS^2}{n-1}\right) = \frac{\sum_{1}^{n}\left(x_i - \bar{x}\right)^2}{n-1}$$

The quantity $\frac{\sum_{1}^{n}(x_i-\bar{x})^2}{n-1}$ is an unbiased estimator of σ^2.

Example Problem 1 (Computing mean and variance). Assume the following data are the average delay at an intersection.

Data: 31, 29, 36, 33, 28, 35, 30, 34, 32, 37

Find the mean and variance:

(a) Sample mean: $\bar{x}$ = (sum x_i)/n = 325/10 = 32.500.
(b) Variance using divisor n: $s^2_n = \Sigma(x_i - \bar{x})^2/n$ = 82.500/10 = 8.250.
(c) Unbiased variance (divisor n – 1): $s^2 = \Sigma(x_i - \bar{x})^2/(n-1)$ = 82.500/9 = 9.167.

(d) The $(n-1)$ divisor produces a slightly larger value because it corrects the downward bias that occurs when $\bar{x}$ is estimated from the same sample.

Example Problem 2 (Effect of an Outlier). Consider if the last value was not 37 but was 50.

Modified data: 31, 29, 36, 33, 28, 35, 30, 34, 32, 50

What are the effects of this outlier?

(a) New mean: $\bar{x} = 338/10 = 33.800$.
New unbiased variance: $s^2 = 351.600/9 = 39.067$.

(b) The outlier increases variance (and standard deviation) and, therefore, increases the confidence-interval half-width ($\varepsilon = t \cdot s/\sqrt{n}$ or $z \cdot s/\sqrt{n}$). Larger variance also increases the required sample size n (since $n \propto s^2$).

9.4 Confidence Level and Confidence Interval

Let Z be the random variable $\left[(\overline{x}-\mu)/\sqrt{\sigma^2/n}\right]$ and have a standard normal distribution with a mean of zero and a standard deviation of one, $Z \sim N(0,1)$. Note that μ is the mean and (σ^2/n) is the variance of $\overline{x}$. Let $z_{(1-\alpha/2)}$ be the $(1-\alpha/2)$ percentile point on the $N(0,1)$ distribution. Then, it can be written that the probability of Z being greater than or equal to $-z_{(1-\alpha/2)}$ and less than or equal to $z_{(1-\alpha/2)}$ is $(1-\alpha)$:

$$P(-z_{(1-\alpha/2)} \leq Z \leq z_{(1-\alpha/2)}) = 1-\alpha$$

Based on the central limit theorem, if the sample size is sufficiently large (usually $n \geq 30$), the distribution of x will be a normal distribution with a mean of $\overline{x}$ and the standard deviation of $s/\sqrt{n}$, i.e. $x \sim N(\overline{x}, s/\sqrt{n})$. This is used to build a confidence interval on the estimation of the mean of the population. The difference between the computed mean and the true mean is $\varepsilon = (\overline{x}-\mu)$ and is called the tolerable error in the estimation of the true mean. The random variable Z which is $[(\overline{x}-\mu)/(s/\sqrt{n}]$ is approximately distributed as a standard normal random variable (Law & Kelton, 2000; Hogg & Tanis, 1977). This property is used to build a confidence interval:

$$P\left(-Z_{(1-\alpha/2)} \leq (\bar{x}-\mu)\Big/\left(s/\sqrt{n}\right) \leq Z_{(1-\alpha/2)}\right) = 1-\alpha$$

It can be written as

$$P\left(-Z_{(1-\alpha/2)} * s/\sqrt{n} \leq (\bar{x}-\mu) \leq Z_{(1-\alpha/2)} * s/\sqrt{n}\right) = 1-\alpha$$

Then, it is simplified to

$$P\left(\bar{x} - z_{(1-\alpha/2)} * s/\sqrt{n} \leq \mu \leq \bar{x} + z_{(1-\alpha/2)} * s/\sqrt{n}\right) = 1-\alpha$$

The confidence interval is the distance between the lower and upper limits. Here, the lower limit of the true mean of the population is $(\bar{x} - z_{(1-\alpha/2)} * s/\sqrt{n})$ and the upper limit of it is $(\bar{x} + z_{(1-\alpha/2)} * s/\sqrt{n})$ with $100(1-\alpha)$ confidence level.

When n is small (less than 30), the random variable $(\bar{x}-\mu)/(s/\sqrt{n})$ would have a "t" distribution with $n-1$ degrees of freedom (df) (Law & Kelton, 2000). So, the confidence interval will be

$$P(\bar{x} - t_{(n-1,\ 1-\alpha/2)} * s/\sqrt{n} \leq \mu \leq \bar{x} + t_{(n-1,\ 1-\alpha/2)} * s/\sqrt{n}) = 1-\alpha$$

where $t_{(n-1,\ 1-\alpha/2)}$ is the upper $1-\alpha/2$ critical point for the t-distribution with $n-1$ df. It should be noted that the confidence obtained based on the t-distribution is wider than the one from the normal distribution because $t_{(n-1,\ 1-\alpha/2)} > z_{(1-\alpha/2)}$ (Law & Kelton, 2000). As n gets larger, the difference gets smaller.

Example Problem 3 (Data of Paired *t*-test). Let's find a confidence interval for the example problem we used in the paired t-test. Recall that in that example, the mean $(\bar{x})$ was –0.15, variance (s^2) was 2.356111, sample size (n) was 10, and t-table for 9 degrees of freedom and 95% confidence level in two-sided test $(t_{(9,\ 0.975)})$ was 2.262:

$$P\left(-0.15 - 2.262 * \sqrt{2.356111}/\sqrt{10} \leq \mu \leq -0.15 + 2.262\right.$$

$$*\sqrt{2.356111}\,/\,\sqrt{10}\Big) = 1 - 0.05$$

$$P\,(-0.15 - 1.098 \le \mu \le -0.15 + 1.098) = 0.95$$

$$P\,(-1.248 \le \mu \le 0.948) = 0.95$$

Thus, with 95% confidence, the lower limit of the population mean will be −1.248 and the upper limit of the population mean will be 0.948.

Example Problem 4 (Confidence Interval). Given $\bar{x} = 32.500$, $s = \sqrt{s^2} = \sqrt{9.167} = 3.028$, $n = 10$, df $= 9$, $t_9{,}0.975 = 2.262$:

$$\text{Half-width} = t{\cdot}s/\sqrt{n} = 2.262 * 3.028/\sqrt{10} = 2.166.$$
$$95\%\ \text{CI: } (\bar{x} - \text{half}, \bar{x} + \text{half}) = (30.334, 34.666).$$

Interpretation: With 95% confidence, the true mean delay μ lies between the lower and upper bounds.

Example Problem 5 (Effect of Confidence Level). Compute CI for 90% and 99%.

90% CI uses $t_9{,}0.95 = 1.833$: half-width $= 1.755 \rightarrow$ CI $= (30.745, 34.255)$
99% CI uses $t_9{,}0.995 = 3.25$: half-width $= 3.112 \rightarrow$ CI $= (29.388, 35.612)$

Higher confidence requires a larger critical value, producing a wider interval.

9.5 Determining Sample Size

To determine the sample size (it may be the number of replications of the simulation model), a decision about the tolerable error "ε" needs to be made. This can be done either by assigning a predetermined value to it or by expressing it as a percentage of the mean value of the performance measure. Then, one needs to know the value of the standard deviation "*s*" to determine the number of replications of the simulation model.

The sample size (number of replications of the simulation run) can be decided by one of the two ways:

1. fixed sample size (pre-determined number);
2. sequential procedure.

9.6 Fixed Sample Size

The discussion above about confidence interval is used to determine sample size (Sokolowski & Banks, 2010). We want the absolute value of the difference between the computed mean (sample mean) and the population mean (true mean), $|(\bar{x} - \mu)|$, to be less than or equal to $(z_{(1-\alpha/2)} * s/\sqrt{n})$ with $100(1 - \alpha)$ confidence level:

$$P(|\bar{x} - \mu|) \leq \left(z_{(1-\alpha/2)} * s / \sqrt{n}\right) = 1 - \alpha$$

$$P(|\varepsilon|) \leq \left(z_{(1-\alpha/2)} * s / \sqrt{n}\right) = 1 - \alpha$$

If it is desirable to have an error that is less than or equal to a predefined value of "ε", then the sample size (n) must be at least equal to

$$n = (z^2_{(1-\alpha/2)})\ s^2/\varepsilon^2$$

The difficulty in using a fixed sample size is that one needs to know "s". When "n" and "s" are known, one can compute the resulting "ε" value.

The main advantage of the fixed sample method is knowing the precision level, "ε", of the estimated true mean. The disadvantage of it is that the user does not have control over the precision level, thus no control over the confidence interval (Law & Kelton, 2000).

Example Problem 6 (Fixed Sample Size). Let's assume you are doing a simulation study, and you want to determine the number of replications (sample size). First, we need to decide what the tolerable error (ε) is. Let's assume $\varepsilon = 0.6$. We also need to know the variance. Let's assume from the previous study we know it is 2.4. The confidence level we'll use is 5%. The Z value from a normal distribution table for 5% confidence level is 1.645. Substituting the known values in the equation yields

$$n = (1.645^2) * 2.4/(0.6^2) = 18.04 \approx 18$$

So, 18 replications of the simulation model are needed. One may ask what if I don't know the variance that should be used. Then, the sequential procedure described in the following section should be used.

9.7 Sequential Procedure

Another way of determining the sample size is the sequential procedure, where the user has control over the precision level (Law & Kelton, 2000). Here, the precision level, (ε), is not known and will be sequentially computed from the following equation until the desired tolerance level is reached:

$$\varepsilon^2 = (z^2_{(1-\alpha/2)})\, s^2/n$$

$$\varepsilon = \left(z_{(1-\alpha/2)} * s\right) / \sqrt{n}$$

The following steps should be taken in the sequential procedure.

Step 1. Make n_0 replications of the simulation as a starting point (may start with several or more runs).
Step 2. Compute $\overline{x}$ and s^2.
Step 3. Compute the precision level "ε" using the values in Steps 1 and 2.
Step 4. If the desired precision level is not reached, make additional replications (you may make one or more replications at this stage) and repeat Steps 2 and 3. If the desired "ε" is obtained, then you have found the appropriate n.

Example Problem 7 (Sample Size for Sequential Procedure). Assume the tolerable error is 15% of the average value of the simulation outcomes ($\varepsilon = 0.15\overline{x}$). What is the sample size to achieve this level of error? Let's start with seven simulation replications. Find the mean and variance for those seven runs: mean = 2.1 and variance = 1.8. Then, compute ε:

$$\varepsilon = \left(z_{(1-a/2)} * s\right) / \sqrt{n} = 1.645 * \sqrt{1.8} / \sqrt{7} = 0.8342$$

Comparing 0.8342 to 0.315 (0.15 * 2.1 = 0.315) indicates that the number of replications should be increased. We make 5 more replications and find the mean and variance for the 12 runs. Now, mean = 2.2 and variance is 1.7. Compute the tolerable error:

$$\varepsilon = \left(z_{(1-a/2)} * s\right) / \sqrt{n} = 1.645 * \sqrt{1.7} / \sqrt{12} = 0.6192$$

Comparing 0.6192 to 0.33 (0.15 * 2.2 = 0.33) indicates that the number of replications should be increased. Now, we make 10 additional replications and find the mean and variance for the 22 replications. The mean is 2.3 and the variance is 1.6. Compute the tolerable error:

$$\varepsilon = \left(z_{(1-a/2)} * s\right) / \sqrt{n} = 1.645 * \sqrt{1.6} / \sqrt{22} = 0.4463$$

Comparing 0.4436 with 0.345 (0.15 * 2.3 = 0.345) indicates that the number of replications should be increased. Make 15 additional replications and find the mean and variance for the 37 replications. The mean is 2.24 and the variance is 1.64. Compute the tolerable error:

$$\varepsilon = \left(z_{(1-a/2)} * s\right) / \sqrt{n} = 1.645 * \sqrt{1.64} / \sqrt{37} = 0.3463$$

Comparing 0.3463 with 0.336 (0.15 * 2.24 = 0.336) indicates that the number of replications should still be increased. We make 5 additional replications and find the mean and variance for the 42 replications. The mean is 2.27 and the variance is 1.68. Compute the tolerable error:

$$\varepsilon = \left(z_{(1-a/2)} * s\right) / \sqrt{n} = 1.645 * \sqrt{1.68} / \sqrt{42} = 0.329$$

Comparing 0.329 with 0.341 (0.15 * 2.27 = 0.341) indicates that 42 replications provide the tolerable error we wanted.

9.8 Quick Alternative to Find Magnitude of *n*

An alternative way of quickly estimating the magnitude of the number of replications is using the following equation with the initial values we started with (or subsequent values). This gives us the magnitude of n:

$$n = (z^2_{(1-\alpha/2)})\ s^2/\varepsilon^2$$

$$n = (1.645^2)\ 1.8/(0.15(2.1))^2 = 49$$

This quickly gives us the idea that the number of replications will be in the 40s. The sequential procedure gave us 42 runs.

Example Problem 8 (Sample Size with fixed Tolerance). What is the sample size if the variance is 2.4 and the desired tolerance error is 0.6?

$$n = z^2\, s^2/\varepsilon^2 = (1.645^2)(2.4)/(0.6^2) = 18.040$$

Required n is at least ceil (18.040) = 19 replications.
It is rounded up because the formula gives the minimum n needed.

Example Problem 9 (Sample Size with Percent-Based Tolerance). Assume the tolerance is 10% of the mean and the standard deviation is 8:

$$\varepsilon = 0.10(\bar{x}) = 0.10 * 40 = 4.000$$

$$n = z^2\, s^2/\varepsilon^2 = (1.96^2)(64)/(4.000^2) = 15.366 \rightarrow \text{ceil} = 16$$

If tolerance is 5%: $\varepsilon = 2.000 \rightarrow n = 61.466 \rightarrow$ ceil = 62. Halving ε increases n by a factor of about 4 (since $n \,\alpha\, 1/\varepsilon^2$).

Example Problem 10 (Sample Size using Sequential Method). Using the sequential method determines the sample size if the tolerance error is 10% of the mean. We start with 8 and jump to 15 and 25 and 35 to see when the sample size is enough.

n	$\bar{x}$	s^2	s	$\varepsilon = z\, s/\sqrt{n}$	Target 0.10 $\bar{x}$
8	12.000	7.200	2.683	1.561	1.200
15	11.600	6.800	2.608	1.108	1.160
25	11.900	6.400	2.530	0.832	1.190
35	12.100	6.500	2.550	0.709	1.210

The first two sample sizes do not meet the requirement, $\varepsilon \leq 0.10\bar{x}$. So, a sample size of 25 meets the required condition.

Example Problem 11 (Quick Magnitude Estimate). Using $n_0 = 8$, $\bar{x} = 12.000$, $s^2 = 7.200$, $\varepsilon = 0.10\bar{x} = 1.200$:
Estimate the sample size:

$$n \approx z^2\, s^2/\varepsilon^2 = (1.645^2)(7.200)/(1.200^2) = 13.53 \rightarrow \text{magnitude} \approx 14$$

The quick estimation gives 14, while we know even a sample size of 15 is not enough.

9.9 Performance Measures for Comparisons

One challenging issue in comparing the simulation results to field data is what criteria or performance measures should be used. In other words, based on what criteria or performance measures do you determine that the simulation outputs are close enough to the field data? There are many ways of making this assessment, and Hollander and Liu (2008) summarized the performance measures used in the literature (some of them are listed here). They also provided a guideline for calibration and validation of Simulation models.

One or more of the following criteria may be used to make the assessment.

9.9.1 *Error*

This error is simply the difference between the simulation outcome and comparable field data:

$$\text{Error} = F_i - S_i$$

9.9.2 *Mean Error (ME)*

Ideally, the mean error should be close to zero, and it should come from individual differences that are small. Having a mean error of zero that comes from a large positive difference that is offset by a large negative difference is not desirable. The error does not tell us how large the deviation is relative to the expected value of that variable. That is when we look at the percent error:

$$(\text{ME}) = \frac{1}{n}\sum_{1}^{n}(F_i - S_i)$$

9.9.3 *Percent Error (PE)*

In general, smaller percentage errors are desirable; the magnitude of tolerable error would depend on many conditions. For example, one may establish that PE should not exceed, say, 5%. It is advised to look at the frequency of the large PE and see if it is acceptable:

$$(\mathrm{PE}) = \frac{F_i - S_i}{F_i}$$

9.9.4 *Mean Percent Error (MPE)*

Mean percent error is also called mean normalized error (MNE). Looking at ME and MPE together gives a better picture of how much deviation exists. Lower values of MPE along with ME values close to zero are desirable outcomes:

$$\mathrm{MPE} = \frac{1}{n}\sum_{1}^{n} \frac{F_i - S_i}{F_i}$$

9.9.5 *Mean Absolute Error (MAE)*

MAE eliminates the chance of positive errors, canceling the negative errors. MAE shows the magnitude of the error regardless of whether it was over prediction or underprediction:

$$\mathrm{MAE} = \frac{1}{n}\sum_{1}^{n} \left|(F_i - S_i)\right|$$

9.9.6 *Mean Absolute Percent Error (MAPE)*

Absolute value MPE is found:

$$\mathrm{MAPE} = \frac{1}{n}\sum_{1}^{n} \frac{\left|(F_i - S_i)\right|}{F_i}$$

9.9.7 *Root Mean Squared Error (RMSE)*

Square root of the MSE is found:

$$\mathrm{RMSE} = \sqrt{\frac{1}{n}\sum_{1}^{n} (F_i - S_i)^2}$$

9.9.8 *Root Mean Squared Normalized Error (RMSNE)*

Square root of the normalized MSE is found:

$$\text{RMSNE} = \sqrt{\frac{1}{n}\sum_{1}^{n}\left(\frac{|(F_i - S_i)|}{F_i}\right)^2}$$

9.9.9 *GEH (Geoffrey E. Havers) Statistic*

GEH is used by some practitioners, but there is no solid statistical concept behind it, though the form of the equation looks like a chi-square test. It is applied to a pair of simulation and field data at a time. One finds the difference squared divided by the average values of the two outcomes and then finds the square root of it:

$$\text{GEH} = \sqrt{\left(\frac{(F_i - S_i)^2}{(F_i + S_i)/2}\right)}$$

If Geh is less than 5, generally, it is considered a good fit, and the similarity of the results is accepted. If it is between 5 and 10, it is a warning to indicate possible mismatch of simulation results to field data, and if it is larger than 10, the similarity is rejected (Aly & Martono, 2022). If there are many pairs to compare, one finds GEH for each pair, and if it is less than 5 for a very large portion of the compared data (Barcelo in his book gives an example of an agency in Great Britain that uses 85%), it is accepted as a good fit.

9.9.10 *Thiel's Indicator*

Thiel (1961) proposed a set of indices and U is one of them (for the other three indicators, see Barcelo or Theil). Thiel's indicator (sometimes referred to as Thiel inequality), U, is computed from the following equation:

$$U = \frac{\sqrt{\frac{1}{n}\sum_{1}^{n}(F_i - S_i)^2}}{\sqrt{\frac{1}{n}\sum_{1}^{n}(F_i)^2} + \sqrt{\frac{1}{n}\sum_{1}^{n}(S_i)^2}}$$

The U values are between 0 and 1. If U is zero, it indicates a perfect fit; however, if U is equal to 1, it indicates the worst fit. For U values less than or equal to 0.2, the simulated results can be accepted as replicating the field data (Barceló, 2010). The other three indices are given in Thiel (1961) and Barceló (2010).

Example Problem 12 (Checking Errors). In a simulation study, the outcome from each simulation run (S_i) has corresponding field data (F_i). Twelve simulation runs were made, and the following results were obtained.

i	F_i	S_i	$F_i - S_i$	$(F_i - S_i)/F_i$	$\|PE_i\|$	GEH_i
1	900	870	30	0.0333	0.0333	1.008
2	1100	1145	−45	−0.0409	0.0409	1.343
3	1050	980	70	0.0667	0.0667	2.197
4	1200	1260	−60	−0.0500	0.0500	1.711
5	950	910	40	0.0421	0.0421	1.312
6	1000	1015	−15	−0.0150	0.0150	0.473
7	1150	1185	−35	−0.0304	0.0304	1.024
8	980	940	40	0.0408	0.0408	1.291
9	1020	1080	−60	−0.0588	0.0588	1.852
10	1080	1045	35	0.0324	0.0324	1.074
11	1250	1305	−55	−0.0440	0.0440	1.539
12	990	955	35	0.0354	0.0354	1.122

Compute errors:

$$\text{ME} = \frac{1}{n}\sum_{1}^{n}(F_i - S_i) = \frac{-20}{12} = -\ 1.667$$

$$\text{MAE} = \frac{1}{n}\sum_{1}^{n}\left|(F_i - S_i)\right| = \frac{520}{12} = 43.333$$

$$\text{MPE} = \frac{1}{n}\sum_{1}^{n}\frac{F_i - S_i}{F_i} = \frac{0.0116}{12} = 0.0010$$

$$\text{MAPE} = \frac{1}{n}\sum_{1}^{n}\frac{\left|(F_i - S_i)\right|}{F_i} = \frac{0.4898}{12} = 0.0408$$

$$\mathrm{RMSE} = \frac{\sqrt{\frac{1}{n}\sum_{1}^{n}(F_i - S_i)^2 = 2095.833}}{12} = 45.780$$

$$\mathrm{RMSNE} = \sqrt{\frac{1}{n}\sum_{1}^{n}\left(\frac{|(F_i - S_i)|}{F_i}\right)^2} = \frac{0.001834}{12} = 0.04283(4.28\%)$$

Example Problem 13 (Checking the Match). Using the data in the previous example, check if the simulation results match the field data.

Compute GEH values:

GEH < 5 for 12/12 locations = 100.0%. If the 85% criteria are used, the model meets the GEH criterion, indicating that simulation results match field data.

Compute Thiel's U:

$$U = \frac{\sqrt{\frac{1}{n}\sum_{1}^{n}(F_i - S_i)^2}}{\sqrt{\frac{1}{n}\sum_{1}^{n}(F_i)^2} + \sqrt{\frac{1}{n}\sum_{1}^{n}(S_i)^2}} = \frac{45.78027}{(1060.5541 + 1065.9718)} = 0.0215$$

U is close to 0 (perfect fit) and well below 0.2, so the simulation replicates the field data acceptably by this guideline.

Overall, the simulation results very much agree with field data: ME is near zero (−1.667 veh/h), indicating little overall bias; MAE and RMSE are moderate relative to typical volumes (~1000 veh/h). MAPE is about 4.08%, suggesting small typical relative errors. All GEH values are below 5, meeting common acceptance criteria. Thiel's $U \approx 0.021$ indicates very good agreement.

Exercises

Part 1. Questions

1. Explain why simulation outputs vary from run to run.
2. Explain the meaning of a 95% confidence interval in simulation studies.

3. Why does increasing the number of simulation replications increase confidence in the estimated mean?
4. Why is dividing by $(n - 1)$ preferred when estimating population variance?
5. When should the t-distribution be used instead of the normal distribution?
6. What is the main disadvantage of the fixed sample size method or the sequential procedure?
7. Is there any relation between confidence level and confidence interval?
8. When should one use the mean difference between simulation and field data (ME), absolute value of ME, relative magnitude of ME, absolute value of ME, and root mean square of ME?
9. Why is ME alone insufficient for validating simulation models?
10. Explain how increasing the sample size reduces the confidence interval width.

Part 2. Problems

Problem 1. A simulation model is run 10 times, producing the following average delays (seconds): 18.2, 19.5, 17.8, 20.1, 18.9, 19.2, 18.4, 17.9, 19.8, 18.7.

a. Compute the sample mean, the sample variance using the divisor n, and the unbiased variance using divisor $(n - 1)$.
b. Construct a 95% confidence interval for the true mean.

Problem 2. In a simulation study, the average travel time was 42.5 min, the variance was 16, and the sample size was 64.

a. Compute the 99% confidence interval.
b. Compute the tolerable error (ε).
c. Explain how the interval would change if n were reduced to 25.

Problem 3. Assume in a simulation study that the desired absolute error is 0.4 vehicles and the estimated variance is 2.5.

a. Determine the required number of simulation replications.
b. If the desired error is reduced to 0.2, how does the required n change?
c. Explain the relationship between ε and n.

Problem 4. In a simulation study, the desired precision is 12% of the estimated mean. The estimated mean is 30 sec and the variance is 9.

a. Determine the required number of replications.
b. If the variance increases to 16, how does the required n change?
c. Explain why variance strongly affects computational effort.

Problem 5. In a sequential procedure to determine the sample size, the initial run is 8 ($n = 8$), the average is 15, and the variance is 4.

a. Compute current ε.
b. Is precision acceptable?
c. Estimate the required magnitude of n.

Problem 6. Use the following data:

Location	Field (F_i)	Simulation (S_i)
1	400	390
2	520	550
3	610	600
4	480	500
5	450	430
6	700	720
7	560	540
8	630	650
9	510	495
10	580	600

a. Compute the error for each location.
b. Compute ME.
c. Compute MAE.
d. Compute RMSE.
e. Identify which measure is most sensitive to large deviations.
f. Compute PE.
g. Compute MPE.
h. Compute MAPE.
i. Discuss systematic overestimation or underestimation.

Problem 7. Use the data from the previous problem.

a. Compute GEH for each location.
b. Determine how many locations satisfy GEH < 5.
c. Does the model meet the 85% acceptance guideline?
d. Discuss the limitations of GEH.
e. Compute Thiel's *U*.
f. Compare conclusions from RMSE, GEH, and Thiel's *U*.

References

Aly, S. H., Martono, S., & Harusi, N. M. R. (2022). Analysis of motorized vehicles performance at signalized intersections based on micro simulation. *IOP Conference Series Earth and Environmental Science, 1117*(1), 012024.

Barceló, J. (ed.). (2010). *Fundamentals of Traffic Simulation.* Springer, New York.

Hogg, R. V. & Tanis, E.A. (1977). *Probability and Statistical Inference.* Macmillan Pub. Co.

Hogg, R. V., Tanis, E. A., & Zimmerman, D. (2021). *Probability and Statistical Inference* (10th ed.). Pearson, New York, NY.

Hollander, Y. & Liu, R. (2008). The principles of calibrating traffic microsimulation models. *Transportation, 35*, 347–362. DOI: 10.1007/s11116-007-9156-2.

Law, A. M. & Kelton, W. D. (2000). *Simulation Modeling and Analysis* (3rd ed.). McGraw-Hill Inc., New York, NY.

Snedecor, G. W. & Cochran, W. G. (1989). *Statistical Methods* (8th ed.). Iowa State University Press, Ames.

Sokolowski, J. A. & Banks, C. M. (Ed.) (2010). *Modeling and Simulation Fundamentals: Theoretical Underpinnings and Practical Domains*. Wiley & Sons, New Jersey.

Thiel, H. (1961). *Economic Forecasts and Policy.* North-Holland Publishing Company, Amsterdam.

Glossary and Abbreviations

AADT (Annual Average Daily Traffic)—Average daily traffic volume computed from continuous counts collected over at least one year.

AAWT (Annual Average Weekday Traffic)—Average weekday traffic volume calculated over a year.

Accreditation—A non-technical decision process that certifies a simulation model is acceptable for a specific application.

Acceleration/Deceleration Pattern—The change in vehicle speed over time, often used in microscopic validation.

ADT (Average Daily Traffic)—Average number of vehicles passing a point on a typical day, usually estimated from short-term counts.

Analytical Model—A mathematical representation of a system using equations that can be solved without simulation.

Antithetic Variates (AV)—A variance reduction technique that induces negative correlation between paired simulation runs to reduce output variability.

Arrival Volume—Number of vehicles arriving upstream of a bottleneck or queue during a given time period.

Autocorrelation—Correlation between values of the same variable at different time lags in a time series.

Average Headway (h)—Average time difference between successive vehicles passing a point.

Average Spacing (d)—Average distance between successive vehicles measured from front bumper to front bumper.

AWT (Average Weekday Traffic)—Average traffic volume for weekdays only.

Batch Means Method—A technique for analyzing simulation output by dividing one long simulation run into several batches to estimate variance and confidence intervals.

Benchmark Data—Reliable reference data (e.g. field data) used to validate simulation outputs.

Calibration—The process of adjusting simulation model parameters so that model outputs closely match real-world data.

Calibration, Verification, and Validation (CVV)—Three important stages in making sure simulation results are valid.

Capacity—Maximum sustainable hourly flow rate under prevailing roadway, traffic, and control conditions.

Car Following (CF) Model—A mathematical model describing how a following vehicle responds to the movement of a leading vehicle.

Cell Transmission Model (CTM)—A discretized macroscopic traffic model that divides a roadway into cells and updates flow between cells over time.

Cellular Automata (CA) Model—A traffic simulation approach in which the roadway is divided into discrete cells that are either occupied or empty, and vehicles move according to predefined rules.

Chi-Square (χ^2) Goodness-of-Fit Test—A statistical test comparing observed and expected frequencies to assess similarity between distributions.

Cluster Model—A mesoscopic modeling approach where vehicles are grouped into clusters or platoons sharing common properties.

Common Random Numbers (CRN)—A variance reduction technique that uses the same random number streams across different simulation scenarios.

Conceptual Model—A non-computational representation (e.g. flowchart or diagram) of a system's structure and logic before implementation in software.

Confidence Interval—A range within which the true population parameter is expected to lie with a specified probability.

Confidence Level—The probability that a confidence interval contains the true population parameter.

Congestion—A condition occurring when traffic demand exceeds roadway capacity, resulting in queue formation.

Continuity Equation—A conservation equation expressing that the number of vehicles entering and leaving a road section determines the change in density over time.

Continuous Simulation—A simulation in which system state variables change continuously over time.

Correlation Coefficient (r)—A measure of the linear relationship between two variables.

Critical Density (k_c)—Density at which maximum flow (capacity) occurs.

Critical Speed (u_c)—Speed at which maximum flow occurs.

Cumulative Distribution Function (CDF)—A function representing the probability that a random variable is less than or equal to a given value.

Demand—Total number of vehicles desiring to pass through a bottleneck, including arrivals and queued vehicles.

Density (*D* or *k*)—Number of vehicles occupying a unit length of roadway.

Design Hour Volume (DHV)—Traffic volume used for roadway design.

Deterministic Simulation—Simulation in which all input parameters are fixed and no randomness exists.

Discrete-Event Simulation (DES)—Simulation where system state changes only at specific events.

Dynamic V&V Methods—Verification and validation methods that evaluate simulation output by running the model.

Edie Model—A nonlinear two-regime speed–density model.

Experimental Design—Process of planning simulation runs to evaluate system performance.

External Variability—Changes in simulation output caused by changes in input parameters.

F-Test—A statistical test used to compare the variances of two independent samples.

Free Flow Speed (u_f)—Speed of vehicles when traffic density approaches zero.

Fundamental Relationship of Traffic Flow—Relationship among flow, density, and speed. $q = k \times u$.

Gap—Distance or time between vehicles excluding vehicle length.

GEH Statistic—A goodness-of-fit measure used to compare simulated and observed volumes.

GM (Gazis–Herman–Rothery) Model—A generalized car following model.

Greenberg Model—A logarithmic speed–density model.

Greenshields Model—A linear speed–density model.

Harmonic Mean—Averaging method used to compute Space Mean Speed (SMS).

Headway—Time interval between successive vehicles passing a point.

Highway Capacity Manual (HCM) Model—Empirical speed-flow relationship used for capacity analysis.

Hysteresis—Looped relationship occurring during traffic breakdown and recovery.

Independent and Identically Distributed (IID)—Random variables that are mutually independent and have the same distribution.

Initial Transient State (Warm-Up Period)—Period at the beginning of simulation during which outputs may not represent steady state.

Internal Variability—Variability in simulation output caused by stochastic elements.

Jam Density (k_j)—Maximum possible traffic density when vehicles are stopped.

Kolmogorov–Smirnov (K–S) Test—A non-parametric goodness-of-fit test.

Lag (k)—Separation between observations in time series analysis.

Lane-Changing (LC) Model—A logic model describing how vehicles change lanes.

Linear Congruential Generator (LCG)—A pseudo-random number generator algorithm.

Macroscopic Model—Traffic model using aggregate variables.

Mean Absolute Error (MAE)—Average of absolute differences between simulated and observed values.

Mean Absolute Percentage Error (MAPE)—Average of absolute percentage differences.

Mean Error (ME)—Average difference between simulated and observed values.

Mean Percent Error (MPE)—Average percentage difference between simulated and observed values.

Mesoscopic Model—Hybrid modeling approach combining macro and micro elements.

Microscopic Model—Traffic model representing individual vehicles.

Microscopic Validation—Validation performed at the individual vehicle level.

Mixed LCG—Linear congruential generator with nonzero increment.

Model Validation—Process of determining whether a simulation model accurately represents reality.

Model Verification—Process of ensuring that the model is correctly implemented.

Monte Carlo Simulation—Static simulation method using repeated random sampling.

Moving Queue (Dynamic Queue)—Queue in which vehicles move slowly rather than remain stopped.

Operational Validation—Testing whether a model accurately represents system behavior under intended conditions.

Paired Sample *t*-Test—Statistical test comparing paired observations.

Peak Hour Factor (PHF)—Measure of traffic volume variation within the peak hour.

Percent Error (PE)—Relative difference between simulated and observed values.

Period (RNG)—Length of sequence before random numbers repeat.

Platoon—Group of vehicles traveling together.

Point Speed—Speed measured at a single location.

Pooled Variance—Combined estimate of variance from two populations.

Queue—Accumulation of vehicles when demand exceeds capacity.

Random Number—Numeric value generated by an RNG.

Random Number Generator (RNG)—Algorithm that produces pseudo-random numbers.

Random Variable—Function assigning numerical values to random outcomes.

Random Variate—Specific value drawn from a probability distribution.

Reaction Time (T)—Time delay between stimulus and driver response.

Regression Analysis—Statistical method for modeling relationships between variables.

Replication—Independent simulation run with different random seeds.

Replication Method—Approach to handle internal variability via multiple runs.

Root Mean Squared Error (RMSE)—Square root of the mean squared difference between simulated and observed values.

Section Speed—Average speed measured over a roadway segment.

Sensitivity Analysis—Method to determine the effect of input changes on output.

Sequential Sampling Procedure—Method for determining sample size incrementally.

Simulation—Modeling tool used to analyze complex systems.

Space Mean Speed (SMS)—Average speed over a road segment (harmonic mean).

Speed (u)—Rate of vehicle travel.

Speed Profile—Graph of vehicle speed over time.

Spacing—Distance between front bumpers of successive vehicles.

Static Queue—Queue in which vehicles are stopped.

Static V&V Methods—Verification methods examining model structure without execution.

Statistical Validation—Objective comparison of simulation and field data using statistical tests.

Stochastic Simulation—Simulation incorporating randomness.

Student's t-Test—Statistical test comparing means when variance is unknown.

Theil's Inequality Coefficient (U)—Measure of forecasting accuracy.

Time Mean Speed (TMS)—Arithmetic mean of spot speeds.

Time Series Analysis—Statistical analysis of data collected over time.

Traffic Modeling and Simulation (TMS)—Application of modeling and simulation techniques to traffic systems.

Trajectory Plot—Graph of vehicle position vs time.

Underwood Model—Exponential speed–density model.

Unbiased Estimator—Estimator whose expected value equals the true parameter.

Variance Reduction Techniques—Methods to decrease simulation output variability.

Verification, Validation, and Accreditation (VVA)—Processes ensuring correctness, accuracy, and suitability of a model.

Volume—Number of vehicles passing a point during a specified time.

Welch's *t*-Test—Two-sample *t*-test for unequal variances.

Work Zone—Roadway area with reduced capacity due to construction.

Z-Test—Statistical test used when population variance is known or the sample size is large.

Index

www.ingramcontent.com/pod-product-compliance
Lightning Source LLC
LaVergne TN
LVHW021133160826
845679LV00016B/1729

* 9 7 8 9 8 1 1 2 8 7 3 0 5 *